# Studies in Applied Philosophy, Epistemology and Rational Ethics

Volume 38

**Series editor**

Lorenzo Magnani, University of Pavia, Pavia, Italy
e-mail: lmagnani@unipv.it

**Editorial Board**

Atocha Aliseda
Universidad Nacional Autónoma de México (UNAM), Coyoacan, Mexico

Giuseppe Longo
Centre Cavaillès, CNRS—Ecole Normale Supérieure, Paris, France

Chris Sinha
School of Foreign Languages, Hunan University, Changsha, P.R. China

Paul Thagard
Waterloo University, Waterloo, ON, Canada

John Woods
University of British Columbia, Vancouver, BC, Canada

## About this Series

Studies in Applied Philosophy, Epistemology and Rational Ethics (SAPERE) publishes new developments and advances in all the fields of philosophy, epistemology, and ethics, bringing them together with a cluster of scientific disciplines and technological outcomes: from computer science to life sciences, from economics, law, and education to engineering, logic, and mathematics, from medicine to physics, human sciences, and politics. It aims at covering all the challenging philosophical and ethical themes of contemporary society, making them appropriately applicable to contemporary theoretical, methodological, and practical problems, impasses, controversies, and conflicts. The series includes monographs, lecture notes, selected contributions from specialized conferences and workshops as well as selected Ph.D. theses.

## Advisory Board

A. Abe, Chiba, Japan
H. Andersen, Copenhagen, Denmark
O. Bueno, Coral Gables, USA
S. Chandrasekharan, Mumbai, India
M. Dascal, Tel Aviv, Israel
G.D. Crnkovic, Västerås, Sweden
M. Ghins, Lovain-la-Neuve, Belgium
M. Guarini, Windsor, Canada
R. Gudwin, Campinas, Brazil
A. Heeffer, Ghent, Belgium
M. Hildebrandt, Rotterdam,
 The Netherlands
K.E. Himma, Seattle, USA
M. Hoffmann, Atlanta, USA
P. Li, Guangzhou, P.R. China
G. Minnameier, Frankfurt, Germany
M. Morrison, Toronto, Canada
Y. Ohsawa, Tokyo, Japan
S. Paavola, Helsinki, Finland
W. Park, Daejeon, South Korea

A. Pereira, São Paulo, Brazil
L.M. Pereira, Caparica, Portugal
A.-V. Pietarinen, Helsinki, Finland
D. Portides, Nicosia, Cyprus
D. Provijn, Ghent, Belgium
J. Queiroz, Juiz de Fora, Brazil
A. Raftopoulos, Nicosia, Cyprus
C. Sakama, Wakayama, Japan
C. Schmidt, Le Mans, France
G. Schurz, Dusseldorf, Germany
N. Schwartz, Buenos Aires, Argentina
C. Shelley, Waterloo, Canada
F. Stjernfelt, Aarhus, Denmark
M. Suarez, Madrid, Spain
J. van den Hoven, Delft,
 The Netherlands
P.-P. Verbeek, Enschede,
 The Netherlands
R. Viale, Milan, Italy
M. Vorms, Paris, France

More information about this series at http://www.springer.com/series/10087

Marta Bertolaso · Nicola Di Stefano
Editors

# The Hand

Perception, Cognition, Action

Springer

*Editors*
Marta Bertolaso
Institute of Philosophy of Scientific
 and Technological Practice
Università Campus Bio-Medico di Roma
Rome
Italy

Nicola Di Stefano
Institute of Philosophy of Scientific
 and Technological Practice
Università Campus Bio-Medico di Roma
Rome
Italy

ISSN 2192-6255 ISSN 2192-6263 (electronic)
Studies in Applied Philosophy, Epistemology and Rational Ethics
ISBN 978-3-319-66880-2 ISBN 978-3-319-66881-9 (eBook)
DOI 10.1007/978-3-319-66881-9

Library of Congress Control Number: 2017951180

© Springer International Publishing AG 2017
This work is subject to copyright. All rights are reserved by the Publisher, whether the whole or part of the material is concerned, specifically the rights of translation, reprinting, reuse of illustrations, recitation, broadcasting, reproduction on microfilms or in any other physical way, and transmission or information storage and retrieval, electronic adaptation, computer software, or by similar or dissimilar methodology now known or hereafter developed.
The use of general descriptive names, registered names, trademarks, service marks, etc. in this publication does not imply, even in the absence of a specific statement, that such names are exempt from the relevant protective laws and regulations and therefore free for general use.
The publisher, the authors and the editors are safe to assume that the advice and information in this book are believed to be true and accurate at the date of publication. Neither the publisher nor the authors or the editors give a warranty, express or implied, with respect to the material contained herein or for any errors or omissions that may have been made. The publisher remains neutral with regard to jurisdictional claims in published maps and institutional affiliations.

Printed on acid-free paper

This Springer imprint is published by Springer Nature
The registered company is Springer International Publishing AG
The registered company address is: Gewerbestrasse 11, 6330 Cham, Switzerland

# Preface

This volume "The Hand. Perception, Cognition, Action" results from the convergence between research programs developed at the University Campus Bio-Medico (Rome, Italy) and international scholars in order to provide a solid interdisciplinary network of knowledge and research questions.

Recent progresses in the field of artificial hand design and surgery have played a crucial role in the path that eventually led to the present volume. The field of study connected to upper limb prosthetic, in fact, is challenging a wide range of disciplines and providing significant insights on hands' biomechanical functions. Developing competitive projects on biorobotics and prosthetics requires unprecedented interdisciplinary efforts and sensitivity that sometimes call into question the conceptual foundations of disciplines. For example, the possibility to restore tactile perception in amputees through neural interfaces stimulates reflections in neuroscience, philosophy of perception, and psychology. Accordingly, the volume offers an overview of the extraordinary role of the hand in human life by taking into consideration different perspectives, from neuroscience and bioengineering to psychology, from anthropology to philosophy and aesthetics.

In the first part—From Action to Cognition, from Cognition to Action—the authors focus on the impact of the use of hand in human neurobiological and psychological development, highlighting the mutual influence of motor and cognitive domains on human development. By introducing the cognitive properties of the mirror neuron system, L. Fogassi (Chapter "The Cognitive Properties of the Motor System and Mirror Neurons") considers the importance of actions performed by non-human primates' upper limb to understand the goal of actions. The same system in humans is mediated by hand and is responsible for higher cognitive functions, such as social communication, imitation, and action understanding. In Chapter "Children's Object Manipulation: A Tool for Knowing the External World and for Communicative Development," V. Focaroli and J. Iverson consider the use of hand in the first stages of child development. Through an accurate review on the acquisition of manipulative skills in infants, they show that this progressive acquisition is an important developmental milestone, thus highlighting the interdependency between object exploration and language development. The relationship between hand and language is addressed by L. Sparaci and V. Volterra (Chapter "Hands Shaping Communication: From Gestures to Signs"), who provide a complete analysis of the role of handshape from a multimodal and embodied perspective, which results in an integrative review of the several perspectives on the communicative function of the hand. F. Taffoni et al. (Chapter "Primates' Propensity to Explore Objects: How Manual Actions Affect Learning in Children and Capuchin Monkeys") compare the impact of objects exploration in children and in non-human primates, showing that spontaneous exploration mediated by the hand has a great biological significance because it allows to discover and learn the relationship between action and effect and to plan goal-directed tasks. Results from an experiment with non-human primates and children are also discussed there. A.L. Ciancio et al. (Chapter "Current Achievements and Future Directions of Hand Prostheses Controlled via Peripheral Nervous System") conclude the first part of the volume, facing the issue of the hand from one of the most advanced fields of research in bioengineering, i.e. neuroprosthetics. They present the state of the art of technological advancements in the use of neural interfaces in order to restore tactile perception in amputees, showing the complexity of the "hand system" and evidencing the strong linkage between perception, cognition, and action.

The second part of the volume considers the role of the hand in human identity and creativity, highlighting the importance of hand use in those activities that profoundly characterize human rationality and identity, such as playing and listening to music, cooking, or caring. M.T. Russo (Chapter "The Human Hand as a Microcosm. A Philosophical Overview on the Hand and Its Role in the Processes of Perception, Action and Cognition") opens the part with a theoretical overview on the various philosophical discussions on the hand. Starting from the Greek debate between Aristotle and Anaxagoras, The author focuses on the controversy between the primacy of touch (i.e., hand) or that of sight (eye) in the characterization of human intelligence and on the consequences that this early debate has in the modern conception of the mind. J.M. Chillon (Chapter *"Ready-to-Hand"* in Heidegger. Philosophy as an Everyday Understanding of the World and the Question Concerning Technology") tackles a traditional question for Heideggerian scholars, i.e., the question of technology, which is properly centered in the notion of *Zuhandenheit* (readiness-to-hand). The relationship between man and the handiness world is thus presented in connection with the notion of technique and modern technology. Chapters "The Therapeutic Hand" and "Cooking and Human Evolution" face two fundamental dimensions of human life essentially mediated by the hand—i.e., caring and cooking. In the world of health care, the hand represents a medium for building an affective (and effective) relationship between caregivers and patients. The chapter by X. Escribano and A. Pérez-Bellmunt (Chapter "The Therapeutic Hand") thus leads to the anthropology of caring as an essential part of the art of healing and caring. Instrumental, cognitive, and pathic at the same time, the therapeutic hand is here presented through its multidimensional features that cooperate for the care and healing of the human person. In Chapter "Cooking and Human Evolution," grounding on the works by Richard Wrangham, M.P. Chirinos stresses the role of cooking in the transition from hominization to humanization. The role of hand specialization in this biocultural practice made our digestive system so different from other primates, thus impacting also on social behavior and habits. The volume ends considering the aesthetic dimension of humans. In Chapter "Essential to Art," S. Castro faces the relationship between aesthetic experience and touch, framing it within the traditional primacy of "intellectual" senses (i.e., sight and hearing) with respect to lower ones (i.e., smell, taste, and touch). Finally, in Chapter "On the Role of the Hand in the Expression of Music," M. Leman et al. provide a review of the role of the hand in the expression of music, focusing on different dimensions of music experience, i.e., music playing, listening, conducting, and learning.

Integrating biological, technological, and philosophical contributions, the volume aims at providing an innovative perspective on the issues related to hand. Though technological development has always offered new opportunities for scientific and social advancements, nowadays technology and biomedicine are coevolving in extraordinary ways, asking philosophy and epistemology to understand the implications of these movements, e.g. the reshape of traditional philosophical concepts such as the opposition between natural and artificial, or basilar notions such as body, hand, or perception. More academic effort is expected in this direction, i.e. the identification of philosophical issues that are emerging within science and that inextricably connect life, technology, and cultural practices, thus emphasizing their mutual interdependencies.

Rome, Italy

Marta Bertolaso
Nicola Di Stefano

# Contents

**Part I  From Action to Cognition, from Cognition to Action**

The Cognitive Properties of the Motor System
and Mirror Neurons .......................................... 3
Leonardo Fogassi

Children's Object Manipulation: A Tool for Knowing
the External World and for Communicative Development ........... 19
Valentina Focaroli and Jana M. Iverson

Hands Shaping Communication: From Gestures to Signs ............ 29
Laura Sparaci and Virginia Volterra

Primates' Propensity to Explore Objects: How Manual Actions
Affect Learning in Children and Capuchin Monkeys ................ 55
Fabrizio Taffoni, Eugenia Polizzi di Sorrentino, Gloria Sabbatini,
Domenico Formica and Valentina Truppa

Current Achievements and Future Directions of Hand Prostheses
Controlled via Peripheral Nervous System ........................ 75
Anna Lisa Ciancio, Francesca Cordella, Klaus-Peter Hoffmann,
Andreas Schneider, Eugenio Guglielmelli and Loredana Zollo

**Part II  The Hand, Human Identity and Creativity**

The Human Hand as a Microcosm. A Philosophical Overview
on the Hand and Its Role in the Processes of Perception, Action,
and Cognition .............................................. 99
Maria Teresa Russo

*Ready-to-Hand* in Heidegger. Philosophy as an Everyday
Understanding of the World and the Question
Concerning Technology ...................................... 115
José Manuel Chillón

**The Therapeutic Hand** ........................................... 127
Xavier Escribano and Albert Pérez-Bellmunt

**Cooking and Human Evolution** ................................. 147
Maria Pia Chirinos

**Essential to Art** .................................................. 163
Sixto J. Castro

**On the Role of the Hand in the Expression of Music** ................ 175
Marc Leman, Luc Nijs and Nicola Di Stefano

# Part I
# From Action to Cognition, from Cognition to Action

# The Cognitive Properties of the Motor System and Mirror Neurons

**Leonardo Fogassi**

**Abstract** According to the traditional view, the motor system of the cerebral cortex has the fundamental role of driving and controlling movement execution. However, the neurophysiological and anatomical data of the last thirty years demonstrated that the main task of the motor cortex is rather that of coding the motor goals. In fact, motor cortex contains a neural storage of motor representations that are used for the sensorimotor transformations necessary for performing goal-directed actions and, at the same time, code important cognitive functions such as space and object representation and recognition of others' behaviour. In this chapter, it will be described first how space coding and object coding are represented in dedicated frontoparietal networks. Then, most of the chapter will be focused on the description of the functional properties of another frontoparietal network, the mirror neuron system. Firstly, the basic and the most recent characteristics of mirror neurons in the monkey will be presented. Secondly, the main features of the mirror neuron system in humans will be described. The last part of this chapter will be concentrated on two social cognitive functions based on the mirror neuron mechanism: imitation and understanding of others' motor intentions.

**Keywords** Motor goal · Parieto-frontal · Visuomotor · Action observation · Intention understanding

## 1 Introduction

The classical notion of the motor system is that of a sector of the nervous system that is devoted to movement execution. Although this is true when looking at its final output, it does not immediately clarify the different levels of organization of this system and of its strong interactions with other components of neural elaboration, such as perception. Furthermore, it is not evident, at first glance, that this part of the nervous system has a strong role in cognition. This latter aspect will be the

---

L. Fogassi (✉)
Dipartimento di Medicina e Chirurgia, Università di Parma, Parma, Italy
e-mail: leonardo.fogassi@unipr.it

main focus of this article, and a substantial part of it will be devoted to the description of the properties and role of the mirror neuron system, that is probably the best example of the neural motor cognitive properties.

The different levels of organization of the motor system in the brain have been well described by behavioural and psychophysical studies (Jeannerod 1988; Rosenbaum et al. 2007) and by many neurophysiological investigations. First of all, it must be emphasized that the main concept coded by the motor system is "action coding", because the action (e.g. eating a piece of apple) is characterized by a final behavioural goal. As well recognized by behavioural experiments, actions are formed by motor chunks (Bernstein 1996), that is by a series of motor acts, i.e. movements aimed to a goal (e.g. grasping a piece of apple, reaching a table). In turn, motor acts are accomplished through "movements" that we will define here as displacements of single joints (e.g. flexion of a finger). Motor acts are organized in specific sequences aimed to achieve a given behavioural goal (the action goal). This sequence organization requires also, at the execution level, a fluent link between one motor act and the next, in order to obtain the so-called motor "melody" (Luria 1973). For the nervous system is economic to code these different movement levels in different anatomical sectors or by different neuronal mechanisms. Thus, actions and motor acts are coded by higher-order cortical areas and movements are coded by primary motor cortex or even by subcortical areas and executed, as final output, by the spinal motor neurons that directly control muscles contraction. In addition, motor neurons belonging to the cortical side of the motor system are also endowed with properties that let emerge cognitive functions.

## 2 The Motor Cortex Codes the Goal of Motor Acts

One of the most important achievements of the last decades in the neurophysiology of the primates motor system is that neurons of the agranular[1] frontal cortex activate when monkeys perform goal-related motor acts, such as reaching an object, grasping it, manipulating it, rather than during the execution of simple movements, such as arm extension or finger flexion (Rizzolatti et al. 1988). In the premotor area F5, some neurons code motor acts at a high level of abstraction, discharging when the monkey grasps food with the hand or the mouth, or even with a tool that it has been instructed to use (Umiltà et al. 2008). Although "goal" coding is typical of premotor cortex, neurons coding the goal of motor acts have been recorded also in primary motor cortex, where, however, "movement" code prevails (Alexander and Crutcher 1990; Kakei et al. 2001; Umiltà et al. 2008).

All these data lead to the conclusion that premotor cortex contains a "storage" of motor representations, to be considered as an "internal motor knowledge", or, using a metaphor, a "motor vocabulary". To have this motor knowledge means that since our birth we are endowed with a powerful tool that enables us to assign meaning to the external world. This capacity relies on the fact that motor representations contained in the motor cortex can be addressed by the elaborated sensory inputs coming from posterior cortical areas through the rich anatomical connections linking posterior parietal and premotor cortex (see Rizzolatti and Luppino 2001). Since these connections are reciprocal, neurons belonging to these two cortical sectors often show similar properties. More specifically, there are several dedicated parieto-frontal circuits that allow the transformation of the sensory inputs into appropriate motor acts. This transformation also provides an automatic attribution of motor meaning to the sensory input. In conclusion, these parietal-premotor circuits can be included in a common "motor system".

## 3 How the Motor System Codes Space and Objects

A first, dedicated, parieto-premotor circuit is that linking the ventral intraparietal area (VIP), lying in the fundus of the intraparietal sulcus, with ventral premotor area F4. Neurons of this latter area are involved in the execution of goal-directed motor acts, such as reaching with the arm, avoiding or approaching with the trunk, or making facial grimaces (Fogassi et al. 1996). These properties are confirmed by electrical microstimulation that elicits from this area arm, neck, or face movements (Gentilucci et al. 1988). One of the most relevant properties of area F4 is the presence of a class of sensory neurons activated by both the application of a tactile stimulus on a body part, say the face, and to the presentation of a 3D visual stimulus in the sector of space near the tactile receptive field. Approaching stimuli are usually the optimal ones. Interestingly, this visual response is limited to the space around the monkey's body, within the reaching space (peripersonal space), and does not depend on eye position; therefore, it has been defined as somatocentric, that is coded in a system of coordinates centred onto body parts. Noteworthy, when the visual stimulus is approached at increasing speed, the 3D visual RF expands. Thus, although phenomenologically the response of these "bimodal" neurons is visual, it has been interpreted as a potential motor representation of one of the motor acts in space coded in this area. Interestingly, in area VIP there are, beyond purely visual neurons, also many neurons with bimodal properties similar to those of F4, but, as compared to the latter, in the former a lower percentage of them code the visual stimulus in somatocentric coordinates (Duhamel et al. 1998). Thus, it is likely that the transformation of the visual input into a motor format is still not complete in VIP, while it is almost completely achieved in F4. Together, the two areas provide a motor coding of space. This is also confirmed by the fact that, although VIP has no direct motor output, its electrical microstimulation with parameters higher than the classical ones can produce movements of the head and the arm.

The second, dedicated, parieto-premotor circuit is that connecting the anterior intraparietal area (AIP) and the premotor area F5. In order to understand the function played by this circuit, it is useful to remember the concept of "affordance", first proposed by the psychologist Gibson (1979). According to him, objects of different shapes can afford different types of interactions. Thus, the affordances represent the physical properties of the objects that suggest a possible interaction.

Note also that the same object can provide several affordances, the choice of which is driven by the context. As a consequence, an object does not provide an individual with a unique perceptual description, but can generate in her/him several "pragmatic" representations. This is confirmed by the presence of a double dissociation, described by Milner and Goodale (1995): The patient D.F., with a lesion to the inferotemporal cortex, was not able to recognize or discriminate objects, but was perfectly capable to grasp them in a correct way. On the contrary, the patient A.T., with a posterior parietal lesion, could normally perceive objects, but was impaired when he had to interact with them (Jeannerod et al. 1994).

The AIP-F5 circuit is responsible for the transformation of object affordances in the appropriate grasping act needed to interact with it. In a series of studies using a paradigm that allowed to separate the visual response to object presentation from grasping execution, it was possible to characterize the properties of F5 and AIP neurons involved in the grasping circuit (Murata et al. 1997, 2000; Raos et al. 2006). First of all, it has been demonstrated that both F5 and AIP contain motor neurons that discharge only during grasping execution both in light and in darkness. Second, in both areas there are visuomotor neurons, responding to the presentation of specific objects and during grasping of them. Third, there is a category of purely visual neurons, responding during grasping in the light, but not in the dark. Many of them respond, specifically, also to pure object presentation (Sakata et al. 1995; Murata et al. 2000). These properties led Jeannerod et al. (1995) to suggest that AIP visual neurons activate for a specific object affordance and send their output to visuomotor F5 neurons and, maybe, also to AIP visuomotor neurons. F5, informed by AIP on object affordances and taken into account the specific context, would choose, from the motor memory, the grip most appropriate for that affordance. Finally, this choice would be sent back to AIP, keeping active the neurons related to the chosen affordance. Interestingly, visuomotor neurons of area F5, named "canonical" neurons, respond also when the monkey must only observe an object, without grasping it, and this response is specific for the type of object. Interestingly, Fluet et al. (2010) demonstrated that when the monkey can grasp the same object with two different grips (precision grip and whole hand grasp) depending on different instructions, the F5 canonical neuron changes its visual response depending on the instruction. Thus, in analogy with what proposed for the VIP-F4 circuit, the visual responses to object presentation of the visuomotor neurons of the AIP-F5 circuit appear to code a motor representation of the object, rather than a pure visual description. This is also confirmed by recent findings showing that when a transparent barrier is interposed between the monkey and the observed object, the visual response of canonical neurons disappears (Bonini et al. 2014a).

Summing up, the two parieto-premotor VIP-F4 and AIP-F5 circuits are endowed with two main functions. First, they allow to transform space location and object properties in the corresponding potential or actual corresponding motor acts, namely reaching/orienting or grasping, respectively. Second, the visual response of the neurons of the premotor node of each circuit assumes a motor format, thus coding space and object in pragmatic terms. This is a first demonstration that cognitive properties emerge from the organization of the motor system.

## 4 Mirror Neurons in the Monkey

### 4.1 Basic Properties

One of the best examples of motor cognition is represented by the mirror neuron system. Mirror neurons are neurons, originally discovered in the macaque ventral premotor area F5, that discharge both when the monkey executes a motor act (e.g. gasping) and when it observes the same, or a similar, motor act performed by another individual, be it another monkey or a human experimenter (Gallese et al. 1996; Rizzolatti et al. 1996a; Ferrari et al. 2003). Subsequently, they have been described also in the inferior parietal lobule (Gallese et al. 2002; Fogassi et al. 2005; Rozzi et al. 2008), more specifically in the area PFG, that is strongly connected with F5 (see Rozzi et al. 2006). The simple presentation of an object or the observation of a person mimicking a motor act is ineffective in activating these neurons. So, in order to respond, they require the observation of an interaction between a biological effectors (hand or mouth) and an object. Part of mirror neurons respond to the observation of only one motor act, while others respond during the observation of two or three motor acts. Among the motor acts effective in driving mirror neurons response, grasping is the most represented one.

Although the visual response of most mirror neurons is invariant with respect to many details of the observed act, some of them show preference for its specific direction, the space sector (left or right) where it is executed and the hand used by the observed agent to interact with the object.

The motor discharge of mirror neurons is indistinguishable from that of purely motor or canonical neurons. This means that in F5, neurons sharing the same motor properties can have different functions. The peculiarity of mirror neurons consists in their visual response. When this response is compared with the motor one, we can categorize these neurons on the basis of the "*congruence*" between the executed and the observed motor act effective in activating them. Based on this categorization, two different sets of mirror neurons have been defined, i.e. "strictly congruent" and "broadly congruent" neurons. The strictly congruent neurons (about one-third of mirror neurons) are neurons in which the executed and observed motor acts match both in terms of the goal (e.g. grasping) and the means to achieve them (e.g. whole hand grip). The other two-thirds (broadly congruent) show a congruence in terms of the goal, but a lower specificity for the observed act or the type of grip during observation as compared to execution (Gallese et al. 1996).

The congruence between the observed and the executed motor act lead to the idea that these neurons perform a "direct matching" between the visual description of another's motor act with one's own motor representation, allowing the observer to understand, automatically, *what* another individual is doing. In other words, the observation of a motor act would determine, in the observer, an automatic retrieval of the motor representation already available within her/his motor vocabulary.

## 4.2 Mirror Neurons and the Understanding of Motor Goals

The capacity of mirror neurons to enable individuals to understand the goal of motor events has been tested in different neurophysiological investigations. Umiltà et al. (2001) demonstrated that when the monkey can see only part of a goal-directed motor act because part of it is hidden behind a screen, half mirror neurons discharge also in this condition. This result demonstrates that mirror neurons can retrieve the motor representation corresponding to the observed motor act, by memorizing the object disappeared behind the screen and reconstructing the missing part of the motor act through a mental reconstruction based on the stored motor representation of that act. In another study, Kohler et al. (2002) showed that when the monkey hears the sound of a motor act (e.g. peanut breaking), many mirror neurons that respond to the observation of this motor act also respond to listening of the sound alone. These so-called audio-visual mirror neurons provide individuals with the capacity to understand the meaning of a motor act presented through different sensory modalities; noteworthy, this feature is also typical of language comprehension.

The two studies described above clearly show that mirror neurons do not simply code the visual feature of an observed act, but the goal of it. In line with this concept, in another experiment, Umiltà et al. (2008) trained monkeys to use pliers for retrieving food out of reach. After training, they recorded from neurons of area F5 during both execution and observation of grasping acts made with the hand or the tool. The results showed that mirror neurons responding to observation of hand grasping activated also when this act was performed with the tool, showing that these neurons are capable of generalizing goal coding also to other non-biological effectors. Interestingly, the observation of a motor act performed with another tool that the monkey was not trained to use elicited a lower discharge (Rochat et al. 2010).

In several recent experiments performed in our laboratory, grasping neurons of area F5 were recorded while monkeys performed a go/no-go task ("execution task") and when they observed an experimenter performing the same task ("observation task"). Interestingly, in both tasks there was a condition (no-go) in which the agent (monkey or experimenter) was required to refrain from acting (defined as "inaction"). The results of one of these studies (Bonini et al. 2014b) showed that there were neurons activated both when the monkey grasped an object and when it refrained from grasping. The same phenomenon was observed in some mirror neurons that discharged not only during action observation, but also, quite surprisingly, when the observed agent refrained from acting. Thus, it has been concluded that in the ventral premotor cortex, a subset of motor and mirror neurons can encode not only the representation of the own or other's motor act, but also its negation.

The mirror matching mechanism implies that a biological input, be it visual or acoustic, is compared, in the same neuron, with the corresponding motor representation normally coded by this neuron. However, one could argue that the activation of this representation could in principle elicit an overt movement, in other words the actual production of the observed motor act. Obviously, this does not occur, very likely because of an inhibition mechanism. An empirical support to this idea has been provided by Kraskov et al. (2009) who investigated the mirror properties of neurons located in F5 that they could identify as corticospinal neurons, that is neurons capable of exciting spinal motor neurons. They found that, among the corticospinal neurons responding to observation of motor acts, 25% showed complete inhibition of discharge during grasping observation, while they strongly discharged during grasping execution. It is possible that this inhibition has an important role in the inhibition of the observer's movement during action observation.

## 4.3 Mirror Neurons Are Involved in Encoding the Details of Motor Acts

When we observe another's motor act, we automatically understand its goal, independently of the many details provided by the observed act. Thus, understanding the goal is the main function of mirror neurons. However, are these neurons also involved in encoding the details of the observed act? Recent studies have shown that this could be the case.

A study by Caggiano et al. (2009) was aimed to verify whether the visual discharge of mirror neurons can be modulated by the space sector (distance from the observer), where the observed act is performed. In this study, a human experimenter grasped an object, in one condition inside the monkey peripersonal space, in the other outside it, and contemporarily, the discharge of F5 mirror neurons was recorded. The results showed that half mirror neurons were modulated by the distance where the grasping was executed. Among them, fifty per cent fired stronger when the experimenter grasped a piece of food within the monkey peripersonal space (within reaching distance), and the other half when the observed motor act was performed in the extrapersonal space (outside reaching distance). It has been suggested that the space-related differential response of mirror neurons represents the link between the comprehension of others' actions and the possibility to socially interact with the agent performing the action.

Another investigation (Caggiano et al. 2011) highlights the capacity of mirror neurons to visually code goal-directed motor acts in a view-dependent way. Classically, the visual response of mirror neurons has been tested by using a naturalistic approach, with the experimenter executing goal-directed motor acts in front of the monkey. In the study of Caggiano et al., instead, the activity of F5 mirror neurons was evaluated by presenting to the monkey movies representing grasping motor acts seen from different perspectives (frontal, lateral, egocentric).

The first result of this experiment is that mirror neurons can also be activated when the observed motor act is presented in a movie, even though the response is generally weaker than in the naturalistic condition. Furthermore, using movies, the discharge of a large number of mirror neurons is modulated by the visual perspective from which a motor act is seen by an observer. In fact, only a quarter of the studied neurons appeared to be view-independent, while the remaining appeared to encode, in equal percentages, the three different perspectives used in this study. These results suggest that beyond encoding the goal of motor acts, mirror neurons can also contribute to provide the details on specific aspects of the observed act. It has been proposed that this latter function could be performed through the feedback connections linking the mirror neurons frontal and parietal areas with the temporal areas encoding the different pictorial views of a motor act (see also below the section on the parieto-frontal mirror circuit).

## *4.4 The Monkey Network Activated by Action Observation*

As it has been described above, mirror neurons have also been recorded in the inferior parietal cortex. This finding was the result of an investigation prompted by the need to conceptually reconstruct the anatomical circuit that allows the building of the mirror system. It was previously reported that in the temporal lobe, namely in the superior temporal sulcus (STS), there are neurons firing during observation of hand grasping (Perrett et al. 1989). However, these neurons do not discharge during active movements. This means that STS is the origin of the higher-order visual input necessary for the formation of mirror neurons in the parieto-frontal circuit. This point has been recently better clarified by an fMRI experiment in monkey complemented by neuroanatomical tracing studies (Nelissen et al. 2011). In the fMRI experiment, monkeys had to observe videos of grasping motor acts performed by a human agent. The results showed activation in the STS, the inferior parietal lobule and the ventral premotor cortex. A ROI analysis on chosen cytoarchitectonic areas of the parietal lobe showed that the areas activated by grasping observation were PFG (located in the inferior parietal lobe convexity) and AIP (located in the intraparietal sulcus) that, in turn, are connected with both ventral premotor area F5 and the region of the STS. These findings constitute a fundamental basis also for understanding the human mirror system.

## 5 The Mirror System in Humans

The mirror matching mechanism has been a fundamental acquisition in the phylogenesis. Thus, a disappearance of such a mechanism in primates evolution would be very surprising. Indeed the first, pioneering studies in humans showed that the mirror system is present in our species. First of all, a transcanic magnetic stimulation (TMS) study (Fadiga et al. 1995) showed that the motor cortex activates during action observation. In this study, while subjects observed an experimenter grasping an object, a TMS pulse was delivered. Motor-evoked potentials (MEPs) were recorded from the hand muscles of the observer, the same he normally uses to execute the observed motor act. The results showed that the magnetic pulse given during observation determined a specific enhancement of the hand MEPs. Interestingly, also observation of meaningless arm movements determined a MEP activation, suggesting that in humans, differently from monkeys, also observation of non-goal-directed movements can elicit motor cortex activation.

Although the TMS study shows a cortical motor activation, its spatial resolution cannot provide a precise anatomical indication of the cortical areas involved during action observation. Two PET studies first showed that observation of grasping acts activates the superior temporal cortex, the inferior parietal lobule and the ventral premotor cortex/inferior frontal gyrus (IFG) (Rizzolatti et al. 1996b; Grafton et al. 1996). Since then, this cortical network has been confirmed by many fMRI studies (see for review Caspers et al. 2010; Molenberghs et al. 2012; Rizzolatti et al. 2014). In addition, it has been shown (Buccino et al. 2001) that the precentral gyrus (the cortical sector in which the motor cortex is located) becomes active, during motor acts observation, in a somatotopic fashion: Observation of mouth motor acts (e.g. biting an apple) activates the inferior frontal gyrus and the inferior sector of ventral premotor cortex, observation of hand motor acts activates also more dorsal aspects of ventral premotor cortex, and observation of foot motor acts (e.g. pressing a pedal) activates the dorsal premotor cortex. A similar, although rougher, somatotopy, was observed also in the inferior parietal cortex. It can be concluded that when we observe a motor act, the activation of the motor system is not generalized, but involves the field containing the representation of the specific effector used by the observed agent, confirming the specificity of the mirror matching mechanism. Finally, the use of techniques with high temporal resolution showed that during action observation, the activation of the IFG followed that of visual cortex and preceded that of the precentral gyrus (Nishitani and Hari 2000).

In humans, for ethical reasons, it is almost impossible to record from single neurons. For this reason, several authors challenged the idea of the existence of single-mirror neurons in humans and tried to verify it using the fMRI repetition-suppression technique (see Grill-Spector et al. 2006). However, these studies gave contradictory results (Dinstein et al. 2007; Chong et al. 2008; Kilner et al. 2009; Lingnau et al. 2009). On the other hand, single-voxel studies in single subjects (see Gazzola and Keysers 2009) indicated that many voxels were shared by action observation and action execution in the premotor and parietal cortex, thus constituting a strong indication for the presence of a clear overlap of the two activations. In addition, Mukamel et al. (2010) could record from single neurons of epileptic patients, implanted with deep electrodes (one of the few conditions in which this type of recording is possible), required to observe and execute reaching–grasping motor acts and facial expressions. They found single neurons responding to both observation and execution in the Supplementary Motor Cortex (SMA) and hippocampus. Although these areas are not the typical "mirror" areas, this study constitutes a further confirmation for the presence of neurons endowed with a mirror matching mechanism.

## 6 Social Cognitive Functions Based on the Mirror Matching Mechanism: Imitation and Intention Understanding

### *6.1 Imitation*

Although the best direct evidence of the presence of a matching mechanism comes from monkey mirror neurons, it is generally assumed that monkeys do not show "true imitation", that is the capacity to immediately reproduce novel movements (Visalberghi and Fragaszy 1990). This leads to the conclusion that the main function of mirror neurons in monkeys is that of understanding the behaviour of other individuals, in order to react in an appropriate way. On the other hand, monkeys show *neonatal imitation* that consists in the reproduction of some facial expressions such as lip smacking or tongue protrusion, only in the very first days from birth (Ferrari et al. 2006), a phenomenon probably related to the need of establishing a bond with the mother. Interestingly, it has been demonstrated with EEG that monkey frontal cortex activates during both observation and reproduction of these facial expressions (Ferrari et al. 2012), but not during the observation of non-biological movements. This result has been considered as evidence of an early activation of the mirror system.

Also in humans, since many years, neonatal imitation of facial expressions has been documented (Meltzoff and Moore 1977). This phenomenon disappears after about two months; many months later intentional imitation manifests. The possibility that imitation, in infants, is underpinned by a mirror mechanism is strongly suggested by several studies, showing that infants as young as six months show a decrease in EEG intensity in the $\alpha$ band (or in the mu rhythm) during both observation and execution of hand motor acts. In human adults, of course, true imitation is clearly present. Iacoboni and his coworkers (1999) (Koski et al. 2002), using a paradigm in which subjects had to observe and imitate simple finger movements, showed, during imitation, an activation of the IFG and of the inferior parietal cortex. They also found an activation of the superior parietal cortex, probably related to the activation of kinesthetic hand representation. In humans, the most relevant function of imitation is expressed during learning by observation and execution, that is usually the easiest way to learn mastering musical instruments, sports, dance and several complex motor tasks. As a premise, it is important to notice that in many instances the behaviour of the observed agent to be reproduced by the observer is not limited to simple movements or even single motor acts, but is formed by a sequence of acts. Thus, the task of the observer is not only to recognize the single motor acts, but also the sequence of motor acts that forms the whole behaviour to be imitated. The idea is that, since motor acts are understood through the mirror neuron system, this latter should be active during both observation and imitation of a complex behaviour. This issue was first tested by Buccino et al. (2004), in an event-related fMRI study in which participants had to observe and then, after a pause, imitate novel guitar chords played by an expert guitarist. The study showed an activation of subjects' inferior parietal lobule, ventral premotor cortex and IFG, that is of the mirror system, during both observation and imitation, but the activation was stronger during the latter condition. Another result of the study indicates that the mirror system is not enough for imitation of this complex behaviour. In fact, during the phase between observation and imitation, when subjects were preparing the programme for correctly reproducing the observed chord, there was an additional strong activation of the middle prefrontal cortex (BA 46) that disappears during imitation. Very likely this activation is related to the mental work required during the period in which the subject must reorganize the single observed motor acts (recognized through the mirror system) in the sequence that he is going to reproduce. Indeed, the activated sector of prefrontal cortex is known to have a crucial role in action planning and sequence construction.

## 6.2 Understanding Others' Intentions

The issue of how we understand others' intention has been addressed by many disciplines, such as philosophy, psychology and also neurophysiology. In the neurophysiological field, several studies tried to investigate the neural bases of intended movements, in most cases referring to execution of simple motor acts (Snyder et al. 1997; Andersen and Buneo 2002). Studies of the last decades (Fogassi et al. 2005; Bonini et al. 2010) tried to assess whether the response of purely motor and mirror neurons of inferior parietal (IPL) and ventral premotor (PMv) cortex is influenced, during execution and observation of a motor act (grasping), by the final goal of the action in which the act is included. The neurons were recorded during: i) a motor task in which the monkey reached and grasped a piece of food located in front of it and brought it to its mouth (Condition 1) or placed it into a container (Condition 2), and ii) a visual task where the monkey simply observed an experimenter performing the same motor task in front of it. The results showed that the majority of motor and mirror neurons in IPL discharged differently according to the intended goal of the action in which grasping was embedded (either grasping to place or to eat). Although the same effect was found in PMv neurons, the result was, in percentage and intensity, weaker than in IPL. These data suggest that the activity of IPL and, to a lesser extent, of PMV grasping neurons codes the intention of the performing agent. Furthermore, mirror neurons that discharged stronger during execution of one of the two actions also discharged stronger during the observation of the same action.

The interpretation of these data is that the motor system is organized in chains of neurons, each dedicated to a specific action goal (intention). When the monkey executes a motor act (e.g. grasping) that belongs to a specific action, the neuron response depends on whether this neuron is included in the chain coding that action. Similarly, when observing an action, mirror neurons coding a specific motor act retrieves the neuronal chain coding a particular final goal, allowing the monkey to understand the intention of the acting individual.

Interestingly, an fMRI study in humans (Iacoboni et al. 2005) showed that when participants automatically understand the intention underlying a specific action in a context, there is a specific activation of the right IFG, that is of a frontal node of the parieto-frontal mirror system. Another, control condition of the same study demonstrated that this activation occurred independently of whether subjects were instructed or not to recognize the intention underlying the observed action. Thus, it is likely that monkey mirror neurons provide the first neural substrate for an automatic understanding of other's intentions, that, in the evolution, paved the way to the more sophisticated aspects of intentional reading present in humans, coming into play when individuals must make a reasoning on what are others' intentions.

# References

Alexander, G. E., & Crutcher, M. D. (1990). Neural representations of the target (goal) of visually guided arm movements in three motor areas of the monkey. *Journal of Neurophysiology, 64,* 164–178.
Andersen, R. A., & Buneo, C. A. (2002). Intentional maps in posterior parietal cortex. *Annual Review of Neuroscience, 25,* 189–220.
Bernstein, N. A. (1996). *Dexterity and its development.* New York, NY: Psychology Press, Taylor and Francis Group.
Bonini, L., Rozzi, S., Ugolotti Serventi, F., Simone, L., Ferrari, P. F., & Fogassi, L. (2010). Ventral premotor and inferior parietal cortices make distinct contribution to action organization and intention understanding. *Cerebral Cortex, 20,* 1372–1385.
Bonini, L., Maranesi, M., Livi, A., Fogassi, L., & Rizzolatti, G. (2014a). Space-dependent representation of objects and other's action in monkey ventral premotor grasping neurons. *Journal of Neuroscience, 34,* 4108–4119.
Bonini, L., Maranesi, M., Livi, A., Fogassi, L., & Rizzolatti, G. (2014b). Ventral premotor neurons encoding representations of action during self and others' inaction. *Current Biology, 24,* 1611–1614.
Buccino, G., Binkofski, F., Fink, G. R., Fadiga, L., Fogassi, L., Gallese, V., et al. (2001). Action observation activates premotor and parietal areas in a somatotopic manner: an fMRI study. *European Journal of Neuroscience, 13,* 400–404.
Buccino, G., Vogt, S., Ritzl, A., Fink, G. R., Zilles, K., Freund, H. J., et al. (2004). Neural circuits underlying imitation of hand actions: an event related fMRI study. *Neuron, 42,* 323–334.
Caggiano, V., Fogassi, L., Rizzolatti, G., Their, P., & Casile, A. (2009). Mirror neurons differentially encode the peripersonal and extrapersonal space of monkeys. *Science, 324,* 403–406.
Caggiano, V., Fogassi, L., Rizzolatti, G., Pomper, J. K., Their, P., Giese, M. A., et al. (2011). View-based encoding of actions in mirror neurons of area F5 in macaque premotor cortex. *Current Biology, 21,* 144–148.
Caspers, S., Zilles, K., Laird, A. R., & Eickhoff, S. B. (2010). ALE meta-analysis of action observation and imitation in the human brain. *Neuroimage, 50,* 1148–1167.
Chong, T. T., Cunnington, R., Williams, M. A., Kanwisher, N., & Mattingley, J. B. (2008). fMRI adaptation reveals mirror neurons in human inferior parietal cortex. *Current Biology, 18,* 1576–1580.
Dinstein, I., Hasson, U., Rubin, N., & Heeger, D. J. (2007). Brain areas selective for both observed and executed movements. *Journal of Neurophysiology, 98,* 1415–1427.
Duhamel, J. R., Colby, C. L., & Goldberg, M. E. (1998). Ventral intraparietal area of the macaque: Congruent visual and somatic response properties. *Journal of Neurophysiology, 79,* 126–136.

Fadiga, L., Fogassi, L., Pavesi, G., & Rizzolatti, G. (1995). Motor facilitation during action observation: a magnetic stimulation study. *Journal of Neurophysiology, 73*, 2608–2611.

Ferrari, P. F., Gallese, V., Rizzolatti, G., & Fogassi, L. (2003). Mirror neurons responding to the observation of ingestive and communicative mouth actions in the monkey ventral premotor cortex. *European Journal of Neuroscience, 17*, 1703–1714.

Ferrari, P. F., Visalberghi, E., Paukner, A., Fogassi, L., Ruggiero, A., & Suomi, S. J. (2006). Neonatal imitation in rhesus macaques. *PLoS Biology, 4*, e302.

Ferrari, P. F., Vanderwert, R. E., Paukner, A., Bower, S., Suomi, S. J., & Fox, N. A. (2012). Distinct EEG amplitude suppression to facial gestures as evidence for a mirror mechanism in newborn monkeys. *Journal of Cognitive Neuroscience, 24*, 1165–1172.

Fluet, M. C., Baumann, M. A., & Scherberger, H. (2010). Context-specific grasp movement representation in macaque ventral premotor cortex. *Journal of Neuroscience, 30*, 15175–15184.

Fogassi, L., Gallese, V., Fadiga, L., Luppino, G., Matelli, M., & Rizzolatti, G. (1996). Coding of peripersonal space in inferior premotor cortex (area F4). *Journal of Neurophysiology, 76*, 141–157.

Fogassi, L., Ferrari, P. F., Gesierich, B., Rozzi, S., Chersi, F., & Rizzolatti, G. (2005). Parietal Lobe: From action organization to intention understanding. *Science, 308*, 662–667.

Gallese, V., Fadiga, L., Fogassi, L., & Rizzolatti, G. (1996). Action recognition in the premotor cortex. *Brain, 119*, 593–603.

Gallese, V., Fadiga, L., Fogassi, L., & Rizzolatti, G. (2002). Action representation and the inferior parietal lobule. In W. Prinz & B. Hommel (Eds.), *Common mechanisms in perception and action: Attention and performance* (pp. 334–355). Oxford: Oxford University Press.

Gazzola, V., & Keysers, C. (2009). The observation and execution of actions share motor and somatosensory voxels in all tested subjects: Single-subject analyses of unsmoothed fMRI data. *Cerebral Cortex, 19*, 1239–1255.

Gentilucci, M., Fogassi, L., Luppino, G., Matelli, M., Camarda, R., & Rizzolatti, G. (1988). Functional organization of inferior area 6 in the macaque monkey: I. Somatotopy and the control of proximal movements. *Experimental Brain Research, 71*, 475–490.

Gibson, J. (1979). *The ecological approach to visual perception*. Boston: Houghton Mifflin Company.

Grafton, S. T., Arbib, M. A., Fadiga, L., & Rizzolatti, G. (1996). Localization of grasp representations in humans by positron emission tomography. 2. Observation compared with imagination. *Experimental Brain Research, 112*, 103–111.

Grill-Spector, K., Henson, R., & Martin, A. (2006). Repetition and the brain: Neural models of stimulus-specific effects. *Trends in Cognitive Sciences, 10*, 14–23.

Iacoboni, M., Woods, R. P., Brass, M., Bekkering, H., Mazziotta, J. C., & Rizzolatti, G. (1999). Cortical mechanisms of human imitation. *Science, 286*, 2526–2528.

Iacoboni, M., Molnar-Szakacs, I., Gallese, V., Buccino, G., Mazziotta, J. C., & Rizzolatti, G. (2005). Grasping the intentions of others with one's own mirror neuron system. *PLoS Biology, 3*, e79.

Jeannerod, M. (1988). *The neural and behavioural organization of goal-directed movements*. Oxford: University Oxford Press.

Jeannerod, M., Decety, J., & Michel, F. (1994). Impairment of grasping movements following a bilateral posterior parietal lesion. *Neuropsychologia, 32*, 369–380.

Jeannerod, M., Arbib, M. A., Rizzolatti, G., & Sakata, H. (1995). Grasping objects: The cortical mechanisms of visuomotor transformation. *Trends Neuroscience, 18*, 314–320.

Kakei, S., Hoffman, D. S., & Strick, P. L. (2001). Direction of action is represented in the ventral premotor cortex. *Nature Neuroscience, 4*, 1020–1025.

Kilner, J. M., Neal, A., Weiskopf, N., Friston, K. J., & Frith, C. D. (2009). Evidence of mirror neurons in human inferior frontal gyrus. *Journal of Neuroscience, 29*, 10153–10159.

Kohler, E., Keysers, C., Umiltà, M. A., Fogassi, L., Gallese, V., & Rizzolatti, G. (2002). Hearing sounds, understanding actions: Action representation in mirror neurons. *Science, 297*, 846–848.

Koski, L., Wohlschläger, A., Bekkering, H., Woods, R. P., Dubeau, M. C., Mazziotta, J. C., et al. (2002). Modulation of motor and premotor activity during imitation of target-directed actions. *Cerebral Cortex, 12*, 847–855.

Kraskov, A., Dancause, N., Quallo, M. M., Shepherd, S., & Lemon, R. N. (2009). Corticospinal neurons in macaque ventral premotor cortex with mirror properties: A Potential mechanism for action suppression? *Neuron, 64*, 922–930.

Lingnau, A., Gesierich, B., & Caramazza, A. (2009). Asymmetric fMRI adaptation reveals no evidence for mirror neurons in humans. *Proceedings of the National Academy of Sciences USA, 106*, 9925–9930.

Luria, A. R. (1973). *The working brain. An introduction to neuropsychology*. London: Penguin.

Meltzoff, A. N., & Moore, M. K. (1977). Imitation of facial and manual gestures by human neonates. *Science, 198*, 75–78.

Milner, A. D., & Goodale, M. A. (1995). *The visual brain in action*. Oxford: Oxford University Press.

Molenberghs, P., Cunnington, R., & Mattingley, J. B. (2012). Brain regions with mirror properties: a meta-analysis of 125 human fMRI studies. *Neuroscience and Biobehavioral Reviews, 36*, 341–349.

Mukamel, R., Ekstrom, A. D., Kaplan, J., Iacoboni, M., & Fried, I. (2010). Single-neuron responses in humans during observation and execution of actions. *Current Biology, 20*, 1–7.

Murata, A., Fadiga, L., Fogassi, L., Gallese, V., Raos, V., & Rizzolatti, G. (1997). Object representation in the ventral premotor cortex (area F5) of the monkey. *Journal of Neurophysiology, 78*, 2226–2230.

Murata, A., Gallese, V., Luppino, G., Kaseda, M., & Sakata, H. (2000). Selectivity for the shape, size, and orientation of objects for grasping in neurons of monkey parietal area AIP. *Journal of Neurophysiology, 83*, 2580–2601.

Nelissen, K., Borra, E., Gerbella, M., Rozzi, S., Luppino, G., Vanduffel, W., et al. (2011). Action observation circuits in the macaque monkey cortex. *Journal of Neuroscience, 31*, 3743–3756.

Nishitani, N., & Hari, R. (2000). Temporal dynamics of cortical representation for action. *Proceedings of the National Academy of Sciences USA, 97*, 913–918.

Perrett, D. I., Harries, M. H., Bevan, R., Thomas, S., Benson, P. J., Mistlin, A. J., et al. (1989). Frameworks of analysis for the neural representation of animate objects and actions. *Journal of Experimental Biology, 146*, 87–113.

Raos, V., Umiltà, M. A., Murata, A., Fogassi, L., & Gallese, V. (2006). Functional properties of grasping-related neurons in the ventral premotor area F5 of the macaque monkey. *Journal of Neurophysiology, 95*, 709–729.

Rizzolatti, G., & Luppino, G. (2001). The cortical motor system. *Neuron, 31*, 889–901.

Rizzolatti, G., Camarda, R., Fogassi, L., Gentilucci, M., Luppino, G., & Matelli, M. (1988). Functional organization of inferior area 6 in the macaque monkey. II. Area F5 and the control of distal movements. *Experimental Brain Research, 71*, 491–507.

Rizzolatti, G., Fadiga, L., Gallese, V., & Fogassi, L. (1996a). Premotor cortex and the recognition of motor actions. *Brain Research. Cognitive Brain Research, 3*, 131–141.

Rizzolatti, G., Fadiga, L., Matelli, M., Bettinardi, V., Paulesu, E., Perani, D., et al. (1996b). Localization of grasp representations in humans by PET: 1. Observation versus execution. *Experimental Brain Research, 111*, 246–252.

Rizzolatti, G., Cattaneo, L., Fabbri-Destro, M., & Rozzi, S. (2014). Cortical mechanisms underlying the organization of goal-directed actions and mirror neuron-based action understanding. *Physiological Reviews, 94*, 655–706.

Rochat, M. J., Caruana, F., Jezzini, A., Escola, L., Intskirveli, I., Grammont, F., et al. (2010). Responses of mirror neurons in area F5 to hand and tool grasping observation. *Experimental Brain Research, 204*, 605–616.

Rosenbaum, D. A., Cohen, R. G., Jax, S. A., Weiss, D. J., & van der Wel R. (2007). The problem of serial order in behavior: Lashley's legacy. *Human Movement Science, 26*, 525–554.

Rozzi, S., Calzavara, R., Belmalih, A., Borra, E., Gregoriou, G. G., Matelli, M., et al. (2006). Cortical connections of the inferior parietal cortical convexity of the macaque monkey. *Cerebral Cortex, 16,* 1389–1417.

Rozzi, S., Ferrari, P. F., Bonini, L., Rizzolatti, G., & Fogassi, L. (2008). Functional organization of inferior parietal lobule convexity in the macaque monkey: electrophysiological characterization of motor, sensory and mirror responses and their correlation with cytoarchitectonic areas. *European Journal of Neuroscience, 28,* 1569–1588.

Sakata, H., Taira, M., Murata, A., & Mine, S. (1995). Neural mechanisms of visual guidance of hand action in the parietal cortex of the monkey. *Cerebral Cortex, 5,* 429–438.

Snyder, L. H., Batista, A. P., & Andersen, R. A. (1997). Coding of intention in the posterior parietal cortex. *Nature, 386,* 167–170.

Umiltà, M. A., Kohler, E., Gallese, V., Fogassi, L., Fadiga, L., Keysers, C., et al. (2001). "I know what you are doing": a neurophysiological study. *Neuron, 32,* 91–101.

Umiltà, M. A., Escola, L., Intskirveli, I., Grammont, F., Rochat, M., Caruana, F., et al. (2008). When pliers become fingers in the monkey motor system. *Proceedings of the National Academy of Sciences USA, 105,* 2209–2213.

Visalberghi, E., & Fragaszy, D. M. (1990). Do monkeys ape? In S. T. Parker & K. R. Gibson (Eds.), *"Language" and intelligence in monkeys and apes* (pp. 247–273). Cambridge: Cambridge University Press.

# Children's Object Manipulation: A Tool for Knowing the External World and for Communicative Development

Valentina Focaroli and Jana M. Iverson

**Abstract** The progressive acquisition of manipulative skill is an important developmental milestone. It provides infants with an increasing set of opportunities for knowing the external world and for acquiring abilities also relevant to other domains, most especially social interaction. The ability to use the hands to grasp and extend an object in a directed fashion toward an interlocutor facilitates the establishment of shared attention. Thus, the progression in manipulative ability can serve as an agent of change, not only for motor development in general, but also for communication. This chapter will consider the progressive acquisition of manipulative skills during development, their significance for knowing the external world and, in particular, their close relation to the communicative development of children.

**Keywords** Motor development · Communicative development · Infants

## 1 Introduction

In recent years, there has been growing scientific interest in the development of motor skills in children, and in particular the importance of motor competencies for general development (e.g., Thelen 1995; Campos et al. 2000; von Hofsten 2007; Karasik et al. 2011; Libertus and Needham 2010). One theme that has emerged from this body of research is that the acquisition of new motor skills (e.g., object manipulation) creates opportunities to acquire and refine abilities that are relevant for learning in other domains (e.g., language development; Iverson 2010). On the one hand, attention to specific features of the object during manipulation prepares

V. Focaroli (✉)
Laboratory of Developmental Neuroscience, Università Campus Bio-Medico di Roma, Rome, Italy
e-mail: v.focaroli@unicampus.it

J.M. Iverson
Department of Psychology, University of Pittsburgh, Pittsburgh, PA, USA

the infant to map specific meanings to specific referents in the process of lexical acquisition. On the other hand, the ability to grasp and extend an object in a directed fashion toward a social partner facilitates the establishment of shared attention to that object. The progressive acquisition of motor skills, the exploration of spaces, and the practice of new activities thus serve as an agent of change for the development, promoting the development of functional actions for the individual's needs (Thelen 2004).

The present chapter considers the development of manual skills and object exploration in children and is organized into two main sections. The first considers the developmental progression of children's manipulation and describes the role of other, crucial skills relevant to manipulation (e.g., postural control). The second section addresses the interdependency between object manipulation and development in the linguistic and social domains, highlighting the impact of multiple motor developments on children's knowledge of the world and sharing of objects with caregivers. In the final section, we discuss data about the influence of object manipulation, which deeply affects the general human development.

## 2 The Development of Object Manipulation

*Reaching and grasping*. Ordinary actions such as object manipulation develop over an extended period of time. Efficient manipulation requires the development of two motor acts: reaching and grasping. These behaviors are expressions of the integration between different sensorimotor systems combining perceptual discrimination of an object located in space and a goal-oriented manual action toward the object (Rochat and Goubet 1995). In order to perform a successful reaching action, infants need to coordinate visual, auditory, and proprioceptive stimuli (Clifton et al. 1994), modifying their movement depending on the goal of the activity. Therefore, the manipulation of an object requires both motor control and motor planning abilities.

Over development, reaching and grasping movements become gradually more coordinated: At 2 months of age, improved head control facilitates this path, and from 4 months, a new developmental phase appears due to the emergence of abilities such as eye–hand coordination and improved trunk and postural control, which provide a stable base for the reaching movement. By 6 months, infants can adjust their reaches as a function of perceived spatial and physical properties of the object, such as its size (von Hofsten and Rönnqvist 1988), its orientation (Lockman et al. 1984), and whether or not it is reachable (Clifton et al. 1991; Field 1976; McKenzie et al. 1993; Yonas and Granmd 1985; Yonas and Hartman 1993).

Like reaching, the ability to grasp an object is refined slowly over time. Infants must learn to start closing the hand before touching the object and not as a post-encounter reaction. Studies have shown that when infants start to acquire this skill, they begin to adapt the orientation of the hand to the orientation of the object. Lockman et al. (1984) presented dowels oriented both horizontally and vertically to

5- and 9-month-olds, in order to determine at what point during the reach hand orientation approximated that of the dowel. Both groups of children approached the dowel with the appropriate final hand orientation at the grasp, but the 9-month-olds did so earlier during the reach, before they had tactile information about the dowel's orientation. Later on, between 9 and 13 months, infants also acquire the ability to adapt the hand depending on the size of the object. According to von Hofsten and Ronnqvist (1988), this ability increases significantly at 13 months of age: Children showed larger distance between the thumb and the index finger (measured during the presentation of objects different in size) for bigger objects than for the smallest ones.

An additional developmental change in grasping involves the switch from power to precision grip (Halverson 1931). When children perform a power grip, the object is tightly closed between the lower part of the fingers and the palm, while in precision grip the object is held between the thumb and one or more fingers. This transition starts at around 20 weeks of age. Both power and precision grips are apparent at 6 months, but the precision grip becomes gradually predominant during the second year of age, also adapting to the object size (Butterworth et al. 1997).

In sum, multiple systems required by action performance become future oriented and integrated during development. When reaching emerges, movements that are initially jerky and take a circuitous path to the object gradually become smoother and more direct (e.g., Thelen et al. 1993), grasping rapidly improves (e.g., Wimmers et al. 1998), and infants begin to manipulate objects for effective examination. By around 13 months, grasping and reaching become integrated skills, constituting a single action.

*Postural control.* The development of movements such as reaching and grasping also require the acquisition of abilities necessary for the management of multisensory information. Among these essential abilities, adjusting body posture during movements to anticipate upcoming events is crucial. In fact, maintaining postural control while performing other behaviors is one of the most important challenges that infants must face. Every performed action creates inertial forces that move the center of gravity of the body, and therefore infants need to stabilize themselves in advance in order to maintain balance. When a force causes an unexpected disturbance, posture is destabilized, but when infants are able to anticipate upcoming destabilization, they can implement control strategies prior to the disturbance. In the case of forces generated by voluntary movements, anticipatory strategies are required both before and during the execution of movements. Thus, postural control is a critical component of the functional execution of reaching.

Fallang and co-authors (2000) studied the interaction between reaching and posture in infants between 4 and 6 months of age. They assessed postural behavior during reaching in supine via reaching kinematics and center of pressure. Results showed that at 6 months, infants passed from the phase in which motor paths are explored without precise adaptation to environmental constraints, to a phase in which they gradually learn to adapt motor activity to the features of the context. The connection of postural activity and reaching performance in supine at 6 months of age, as suggested by the authors, "it may represent the emergence of a finely tuned relationship between posture and reaching performance once an increased flexibility between postural control and the skill of reaching is achieved" (Fallang et al. 2000, pp. 17).

The development of postural control allows also the acquisition of new postures: The progression from supine to sitting posture also affects object exploration. For example, when infants lie supine, arm movements are more effortful and less easily controlled as they must constantly work against gravity to hold an object within the line of sight (Soska and Adolph 2014). When seated, however, the hands and arms are free to move in less biomechanically challenging ways, the upright head position enlarges the field of view and stabilizes gaze, thereby promoting eye–hand coordination (Bertenthal and von Hofsten 1998; Rochat 1992). When infants can sit independently (self-sit), hands no longer needed for support are free to move, and possibilities for object exploration are enhanced (Harbourne et al. 2013). Rochat and Goubet (1995), found that while pre-sitters and new sitters typically use one hand for object exploration and focus on centrally located objects, experienced sitters expand the exploration of space, using both hands to examine both centrally and laterally positioned objects. Furthermore, when self-sitters explore objects, they typically engage in more combined visual-manual behaviors (e.g., looking at objects while rotating or fingering) that provide critical multimodal information about object properties (Soska et al. 2010).

Object-directed reaching, grasping, and postural control underlie increasingly sophisticated object exploration behaviors that yield information about objects in the world and about the effects of the infant's own actions on those objects. In the next section, we review and discuss its relations to the development of communicative and social domains.

## 3 Human Hand and Language: The Impact of Manipulation Skill on Communicative and Social Development

Object exploration via reaching and grasping movements and the knowledge it generates are foundational for the later development of language and communication. For instance, Fagan and Iverson (2007) found infant object mouthing during vocalization to be related to greater variety in consonants, especially supraglottis (e.g., [d]), known to be a reliable predictor of subsequent language growth and delay (e.g., Stoel-Gammon 1992). Ruddy and Bornstein (1982) found a strong positive correlation between object exploration (e.g., fingering, squeezing, banging) at 4 months of age and parent-reported vocabulary at 12 months. This led Ruff and colleagues (1984) to suggest that infants who more frequently engage in object examination have enhanced opportunities to extract information about object categories critical for lexical development.

The relationship between children's handedness and manipulation activity has been investigated by Kotwica and co-authors (2008). They showed that infants aged between 7 and 13 months with a stable hand preference are better multiple object users than infants with an inconsistent hand preference. Authors assessed the infants' handedness by presenting 21 objects to infants, one at a time, and the hand used for the initial grasp of the toy was recorded (the procedure is detailed in Michel et al. 1985). Infants with stable hand-use preferences more readily acquired another object after storing an object than did infants without stable hand-use preferences. This generates new instances with objects and a different experience of the external world that may promote further development of other motor skills and cognitive abilities. Along these lines, Nelson and colleagues (2014) investigated the timing of lateralization for manipulative actions in infancy and the relation with later language acquisition. Studying the relationships between early handedness and advances in language development, they observed that children with a consistent hand preference (measured from 6 to 14 months) scored higher on the language scale of the Bayley when tested at 24 months compared to children who showed an inconsistent hand preference.

The exploration and sharing of objects also foster the development of shared attention that is foundational for the communicative development and for the learning of words (Tomasello and Todd 1983). Ruddy and Bornstein (1982) found that mothers who more frequently draw their child's attention to objects at 4 months of age had children with larger vocabularies at 12 months of age. Early language acquisition happens mainly during the interactions between caregivers and child (Bruner 1981). Bruner (1983) suggested that children learn language in familiar contexts during social exchanges in what he calls "object-play formats."

In this regard, the development of motor abilities creates significant changes in infants' experience of the world, enriching social interactions between infant and caregiver and providing productive opportunities for language learning and social development (Iverson 2010). The development of independent locomotion radically changes experiences with objects and social partners (Karasik et al. 2011). Through locomotion, an infant can bring an object to the caregiver and sharing attention to it with the adult. A child that can walk has a greater range of exploration and greater access to objects, while infants who cannot yet walk have more limited interactions with objects because they are limited to more proximally located objects.

In addition, reaching and manipulating objects are also crucial for the development of gestural skills. In particular, Fischer and Zwaan (2008) highlighted the correlation between the manipulation of objects during actions, children's understanding of the proper use of them, and the development of semantic meanings. Capirci and colleagues (2005) showed that children produced communicative manual actions from 10 months of age, and that in most cases corresponded in meaning with representational gestures that appeared later. They showed that expressive manual actions have a semantic connection with symbolic gestures (Longobardi et al. 2014). Children's gestures and communication can also be influenced by object properties, as demonstrated by Bernardis and colleagues (2008), who showed that during manipulation activities, children regulate their vocalizations depending on the object size. In particular, the repetition of vocalizations was found to increase when a pointing gesture was used to obtain a large object than a small one.

Thus, while gestures mediate the relationship between early motor abilities and later vocabulary, object manipulation affects gesture and communication development. Therefore, the frequency of manipulative activities, which increases due to refinements in reaching, grasping, and postural control, together with greater participation of the caregiver during social interaction, increases opportunities to hear new verbal input, paving the way for the comprehension of objects and actions that are foundational for both gesture and word production (Longobardi et al. 2014).

The progressive acquisition and development of motor skills such as prehension, object manipulation, and locomotion significantly increase children's opportunities to explore the external world which have fundamental implication for communicative development. Children's propensity to use objects and to explore them, using both hand and mouth, provides information not only about objects but also about their vocalizations, confirming that experiences in the motor domain foster development in other domains (Iverson 2010).

# 4 Conclusion

The study of motor development has recently received increasing interest in the field of human developmental research. Motor skills provide children with means to explore the environment and to acquire knowledge about the external world, fostering development in other domains, such as communicative and language skills. In this chapter, we examined the development of the ability to manipulate objects, a fundamental motor skill typically acquired during the first year of life that is foundational for acquiring knowledge about features of objects. Scientific research clearly showed that the motor, perceptual, and cognitive domains are linked and influence one other during development (e.g., Gibson and Pick 2000; Thelen and Smith 1993). The findings reviewed here show that the acquisition of a new motor skill such as reaching, grasping, postural control, or walking has consequences for infants' abilities in the perceptual, cognitive, social, and language domains.

As foreseen by Berkeley (1709) in the early eighteenth century, in order to acquire information about the external world, humans need an extended tactile experience together with the possibility to move and act in the environment. Piaget (1952) confirmed this philosophical assumption, stating that infants gradually understand the features of the objects through their manual actions. In this sense, he stressed the role of motor skill acquisition, conceptualizing it as a facilitator for perceptual and cognitive development, and highlighting the fact that the development of our intelligence is grounded in "doing." Therefore, exploratory actions become crucial since they are performed in order to acquire information that enable individuals to know their surroundings (de Campos et al. 2012).

Far from being a transparent "glass" between the world and the self, the body organizes human experience of the world and models the way to be into the world. Children's being in the world is characterized by performing or trying to perform actions. Infants have a sense of their bodies as entities incorporated in the environment, and they have an early sense of their selves based on perception; Rochat calls it the *ecological self*. The ecological self provides, through early explorations of the body and the environment, the development of sense of self as something different from the external world, thus developing a more explicit awareness of him/herself. In this dynamic process of mutual influence between action and cognition, living beings not only acquire knowledge about the external world but implicitly receive information about the self (Gibson 2014; Gallagher and Zahavi 2013).

# References

Berkeley, G. (1709). *An essay towards a new theory of vision*. Dublin, Aaron Rhames.
Bernardis, P., Bello, A., Pettenati, P., Stefanini, S., & Gentilucci, M. (2008). Manual actions affect vocalizations of infants. *Experimental Brain Research, 184,* 599–603.
Bertenthal, B., & von Hofsten, C. (1998). Eye, head and trunk control: The foundation for manual development. *Neuroscience and Biobehavioral Reviews, 22,* 515–520.
Bruner, J. (1981). The social context of language acquisition. *Language & Communication, 1,* 155–178.
Bruner, J. (1983). The acquisition of pragmatic com-mitments. In R. Golinkoff (Ed.), *The transition from prelinguistic to linguistic communication* (pp. 27–42). Hillsdale, NJ: Erlbaum.
Butterworth, G., Verweil, E., & Hopkins, B. (1997). The development of prehension in infants: Flalverson revisited, (1976). *British Journal of Developmental Psychology, 15,* 223–236.
Campos, J. J., Anderson, D. I., Barbu-Roth, M. A., Hubbard, E. M., Hertenstein, M. J., & Witherington, D. (2000). Travel broadens the mind. *Infancy, 1,* 149–219.
Capirci, O., Contaldo, A., Caselli, M. C., & Volterra, V. (2005). From action to language through gesture: A longitudinal perspective. *Gesture, 5,* 155–177.
Clifton, R. K., Perris, E. E., & Bullinger, A. (1991). Infants' perception of auditory space. *Developmental Psychology, 27,* 187.
Clifton, R. K., Rochat, P., Robin, D. J., & Bertheir, N. E. (1994). Multimodal perception in the control of infant reaching. *Journal of Experimental Psychology: Human Perception and Performance, 20,* 876–886.
de Campos, A. C., Savelsbergh, G. J., & Rocha, N. A. C. F. (2012). What do we know about the atypical development of exploratory actions during infancy? *Research in Developmental Disabilities, 33,* 2228–2235.
Fagan, M. K., & Iverson, J. M. (2007). The influence of mouthing on infant vocalization. *Infancy, 11,* 191–202.
Fallang, B., Saugstad, O. D., & Hadders-Algra, M. (2000). Goal directed reaching and postural control in supine position in healthy infants. *Behavioural Brain Research, 115,* 9–18.
Field, J. (1976). The adjustment of reaching behavior to object distance in early infancy. *Child Development,* 304–308.
Fischer, M. H., & Zwaan, R. A. (2008). Embodied language: A review of the role of the motor system in language comprehension. *The Quarterly Journal of Experimental Psychology, 61,* 825–850.
Gallagher, S., & Zahavi, D. (2013). *The phenomenological mind*. London: Routledge.

Gibson, J. J. (2014). *The ecological approach to visual perception: Classic edition*. Psychology Press.
Gibson, E. J., & Pick, A. D. (2000). *An ecological approach to perceptual learning and development*. USA: Oxford University Press.
Halverson, H. M. (1931). An experimental study of prehension in infancy by means of systematic cinema records. *Genetic Psychology Monographs, 10,* 107–285.
Harbourne, R. T., Lobo, M. A., Karst, G. M., & Galloway, J. C. (2013). Sit happens: Does sitting development perturb reaching development, or vice versa? *Infant Behavior and Development, 36,* 438–450.
Iverson, J. M. (2010). Developing language in a developing body: The relationship between motor development and language development. *Journal of Child Language, 37,* 229–261.
Karasik, L. B., Tamis-LeMonda, C. S., Adolph, K. E. (2011). Transition from crawling to walking and Infants' actions with objects and people. *Child Development, 82,* 1199–1209.
Kotwica, K. A., Ferre, C. L., & Michel, G. F. (2008). Relation of stable hand-use preferences to the development of skill for managing multiple objects from 7 to 13 months of age. *Developmental Psychobiology, 50,* 519–529.
Libertus, K., & Needham, A. (2010). Teach to reach: The effects of active vs. passive reaching experiences on action and perception. *Vision Research, 50,* 2750–2757.
Lockman, J. J., Ashmead, D. H., & Bushnell, E. W. (1984). The development of anticipatory hand orientation during infancy. *Journal of Experimental Child Psychology, 37,* 176–186.
Longobardi, E., Spataro, P., & Rossi-Arnaud, C. (2014). The relationship between motor development, gestures and language production in the second year of life: A mediational analysis. *Infant Behavior and Development, 37,* 1–4.
McKenzie, B. E., Skouteris, H., Day, R. H., Hartman, B., & Yonas, A. (1993). Effective action by infants to contact objects by reaching and leaning. *Child Development, 64,* 415–429.
Michel, G. F., Ovrut, M. R., & Harkins, D. A. (1985). Hand-use preference for reaching and object manipulation in 6- through 13-month-old infants. *Genetic, Social, and General Psychology Monographs, 111,* 409–427.
Nelson, E. L., Campbell, J. M., & Michel, G. F. (2014). Early handedness in infancy predicts language ability in toddlers. *Developmental Psychology, 50,* 809.
Piaget, J. (1952). *The origins of intelligence in children* (vol. 8, no. 5, pp. 18). New York: International Universities Press.
Rochat, P. (1992). Self-sitting and reaching in 5-8 month old infants: The impact of posture and its development on early eye-hand coordination. *Journal of Motor Behavior, 24,* 210–220.
Rochat, P., & Goubet, N. (1995). Development of sitting and reaching in 5- to 6-month-old infants. *Infant Behavior and Development, 18,* 53–68.
Ruddy, M. G., & Bornstein, M. H. (1982). Cognitive correlates of infant attention and maternal stimulation over the first year of life. *Child Development, 53,* 183–188.
Ruff, H. A., McCarton, C., Kurtzburg, D., & Vaughan, H. G. (1984). Preterm infants' manipulative exploration of objects. *Child Development, 55,* 1166–1173.
Soska, K. C., & Adolph, K. E. (2014). Postural position constrains multimodal object exploration in infants. *Infancy, 19,* 138–161.
Soska, K. C., Adolph, K. E., & Johnson, S. P. (2010). Systems in development: Motor skill acquisition facilitates 3D object completion. *Developmental Psychology, 46,* 129–138.
Stoel-Gammon, C. (1992). Prelinguistic vocal development. In C. Ferguson, L. Menn, & C. Stoel-Gammon (Eds.), *Phonological development* (pp. 439–456). Parkton, MD: York Press.
Thelen, E. (1995). Motor development: A new synthesis. *American Psychologist, 50,* 79–95.
Thelen, E. (2004). The central role of action in typical and atypical development: A dynamic systems perspective. In I. J. Stockman (Ed.), *Movement and action in learning and development: Clinical implications for pervasive developmental disorders* (pp. 49–73). San Diego, CA: Elsevier Academic Press.
Thelen, E., Corbetta, D., Kamm, K., Spencer, J. P., Schneider, K., & Zernicke, R. F. (1993). The transition to reaching: Mapping intention to intrinsic dynamics. *Child Development, 64,* 1058–1098.

Thelen, E., & Smith, L. B. (1993). *A dynamic systems approach to the development of cognition and action*. Cambridge, MA: MIT Press.
Tomasello, M., & Todd, J. (1983). Joint attention and lexical acquisition style. *First Language, 4,* 197–211.
von Hofsten, C. (2007). Action in development. *Developmental Science, 10,* 54–60.
von Hofsten, C., & Rönnqvist, L. (1988). Preparation for grasping an object: A developmental study. *Journal of Experimental Psychology. Human Perception and Performance, 14,* 610–621.
Wimmers, R. H., Savelsbergh, G. J. P., Beek, B. J., & Hopkins, B. (1998). Evidence for a phase transition in the early development of prehension. *Developmental Psychobiology, 32,* 235–248.
Yonas, A., & Granmd, C. E. (1985). Reaching as a measure of Infants' spatial perceptlon. In G. Gottlieb, & N. Krasnegor (Eds.). *The measurement of audition and vision in the first year of postnatal life. A methodological overview*. Norwood, NJ: Ablex.
Yonas, A., & Hartman, B. (1993). Perceiving the affordance of contact m 4- and 5-month-old infants. *Child Development, 64,* 298–308.

# Hands Shaping Communication: From Gestures to Signs

Laura Sparaci and Virginia Volterra

**Abstract** The focus of this contribution is on the importance of handshapes and is based on data from several studies coming from different perspectives and research traditions. First, we will analyse the emergence of handshapes in infant prehension and fine motor development. Then, we will consider how and to what extent handshapes have played a relevant role in research on gestures in children. Finally, we will describe different trends in the linguistic analyses of handshapes in child and adult uses of sign languages. Bringing these perspectives together for the first time in a single paper provides a better general understanding of the relevant role of the human hand in shaping communication.

**Keywords** Symbolic development · Grasping · Gestures · Sign language · Handshapes

## 1 Introduction

Starting at birth, our hands are a powerful means by which we explore environments that surround us and communicate with others. However, the ability to use our hands to carry out tasks, whether basic, as in using a spoon to eat, or complex, as in playing a violin, may require practice and, sometimes, lifelong refinement processes. Many of our everyday activities that involve grasping and holding objects and tools use specific handshapes. Compare, for example, how you would grasp a hammer with how you would hold a pen. The production of specific handshapes is not restricted to grasping objects. For example, while speaking we often use handshapes to produce specific gestures (known as *representational*) that

---

L. Sparaci
Division of Psychology and Language Sciences, University College London (UCL), London, UK

L. Sparaci (✉) · V. Volterra
Institute of Cognitive Sciences and Technologies (ISTC),
National Research Council of Italy (CNR), Rome, Italy
e-mail: laura.sparaci@istc.cnr.it

represent objects or actions with objects, in the absence of the object itself. If you think that this is unusual or rare, try describing to a friend approximately how big a cereal bowl should be without referring to numerical values or making an analogy to another object. Did you use your hands in this description? What shape did your hands have? In order to classify and analyse the handshapes that we use with gestures and to understand their role in human communication, researchers have available to them studies on languages that rely primarily on the visuo-manual modality: sign languages used in deaf communities. Handshapes are essential linguistic components of these languages. For example, a sign language linguist might describe a specific handshape as acting as a classifier form that together with a hand movement root constitutes a complex spatial predicate structure (see Fig. 1 for an example).

Throughout this paper, we will highlight the importance of handshapes for different types of cognitive tasks, from grasping objects and tools, to communicating

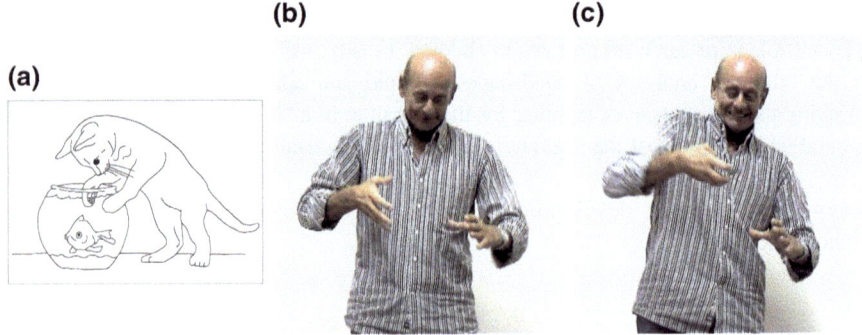

**Fig. 1** Example of case in which to describe a scene in which a cat is trying to grasp a fish in a bowl (A), the *left hand* is used as a classifier for the shape of the fish bowl, and the *right hand* is used as movement root to constitute a complex spatial predicate structure describing the movement of the cat's paw (Adapted from Russo Cardona and Volterra 2007)

**Fig. 2** Different hand areas and approximate object shifts in position made at 3-months, 6-months and 9-months of age

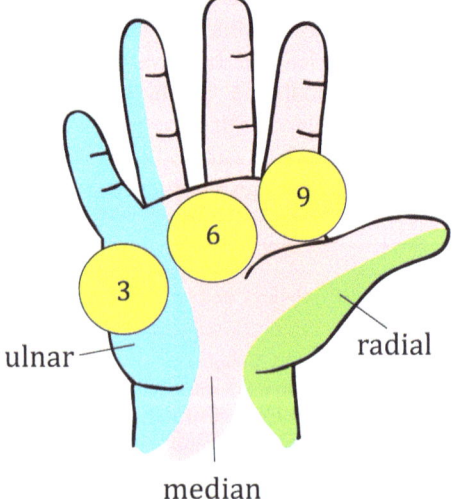

with others through gestures and signs. In doing so, we will describe how different handshapes emerge in development from infant grasping and how they later become essential constituents of communicative and linguistic acts.

While traditional studies on language and communication have focused primarily on speech, in recent years, consistent research has focused on evaluating the possibility of considering language and human communication as *embodied* and *multimodal* (Arbib et al. 2014; Gallagher 2005; Kendon 2004; McNeill 1992; Meteyard et al. 2012). In other words, we are more aware of and clearly acknowledge a phenomenon that was always in plain sight: there is more to language and communication than speech. Our bodies—and in particular our hands—play a fundamental role in our attempts to convey and understand meanings. This has been so for many cultures through the ages. For example, Frank Hamilton Cushing's nineteenth-century pioneering anthropological studies on the Zuni Indians of New Mexico showed how hand use influenced the formation of spoken terms in their language (Cushing 1892).

In this view, language should be considered as a *multimodal* system requiring a multitude of skills (i.e. visual, motoric, auditory, etc.). However, saying that language is embodied, and multimodal does not prove that these skills are *necessary* for conveying content: they might just be emergent behaviours with at most epiphenomenal value. Our main aim is therefore to provide a detailed analysis of the role of handshapes in communication and to attract readers' attention to the rich content that can be conveyed in their use, content that is far from being a simple addendum to speech.

We will show the importance of looking at handshapes ranging from pure motor acts (specifically grasping) and then developing towards symbolic communication in gestures and signs. In other words, the handshapes that we encounter in our everyday gestures and that have been studied in sign languages are grounded in basic embodied motor acts that we acquire in childhood. By beginning with tracing the emergence of handshapes during the development of grasping, we are better able to understand why some handshapes are "easier" than others and how this may affect atypical development. We will then look at the role of handshapes in gesture production and how they contribute to the emergence of symbols and of different gesture types. Finally, we will look into handshapes in sign languages, and how these may be used as conventionalized symbols within a language.

## 2 Developing Grasping in Infancy: The Emergence of Handshapes

As adults, we are rarely conscious of the daily activities in which we grasp, hold and use objects and tools. From when we tightly grasp the handle of our coffee mug in the morning to when we lightly push the light switch at night, our hands work hard to produce a multitude of handshapes that let us make the most of the tools and objects that our cultural backgrounds offer (Wilson 1998). Most of the time, this effort goes unnoticed, but there was a time in our life, in which we had to concentrate and work very hard in order to be able to voluntarily grasp objects and tools in effective ways: infancy.

We do not start from scratch at birth. Hand to mouth movements have been documented in the womb and newborns, and our swinging-from-trees ape ancestors left us with a baggage of reflex mechanisms dedicated to grasping and holding (Butterworth and Hopkins 1988; De Vries et al. 1982; Twitchell 1965). Contrary to voluntary actions, reflex mechanisms are involuntary motor reactions to a stimulus often present at birth or soon afterwards. Reflexes are not "static" and, like our later voluntary grasping movements, they change over time. Thomas Twitchell describes two reflex mechanisms important for later emerging grasping skills and handshapes, which he terms the grasp reflex and the grasp reaction (Twitchell 1965, 1970). The grasp reflex appears around 4 weeks after birth, and in the first stage, it is the act of bringing together the thumb and index finger followed by flexion of other fingers whenever a contact stimulus moves out of the area between the index and thumb. In the following 2–4 weeks, the area affecting this reaction extends to include the entire radial part of the palm (radial extension) and the proximal phalanges (distal extension) and then with time, gradually more fingers become involved.[1]

Around 4- or 5-months, the grasp reflex is followed by the grasp reaction. At this stage, the infant will initially make orienting and groping movements of the hand towards a stimulus. Once again, this reflex changes over time: initially, it will be present only upon stimulation of the radial border of the hand, but after one or two weeks, stimulation of the ulnar side of the hand will lead to groping, orienting and finally grasping a stimulus from the radial side (Twitchell 1965, 1970). These reflex mechanisms set the stage for the first appearance of a variety of handshapes and allow us to observe an essential prelude to the sophisticated changes observed in voluntarily grasping and in how different areas of the hand start to play a role in grasping. Before moving on to the evolution of voluntary grasping; however, a caveat must be presented concerning the theoretical approaches involved in these studies.

Early studies on grasping in infants proceeded from now largely discredited maturational approaches, i.e. the idea that the development of cognitive skills follows a fixed sequence observable at relatively constant chronological ages independent of environmental determinants. For example, Gesell and Ilg's (1937) longitudinal analysis of spoon grasping during feeding between 4- and 13-months of age and Halverson's (1943) description of grasp types using a cube task in the same age span, while providing initial data and fine-grained analyses of grasping in infancy, also maintained the ancillary role of environmental factors. This "internalist" approach focused primarily on the maturation of the nervous system, concentrating on what infants *could do* irrespective of their environment. Later,

---

[1]The hand is commonly divided into different areas that take their names from the nerves that innervate them: radial, median and ulnar. See also Fig. 2 for an approximate mapping of these areas. Proximal areas are areas closer to the palm, while distal areas are closer to the digits.

"externalist" approaches, focusing on what infants *would do* given different contexts for action, demonstrated that organismic, environmental and task-related constraints all play a major role in shaping the emergence of human grasps (Connolly and Elliott 1972; Newell 1986). In our description of grasping development, we try to bring together both worlds by maintaining an embodied perspective, centred neither on inner brain mechanisms nor on external environmental features, but rather on the relationship between the two in order to obtain a more nuanced description of how grasping and hand configurations change in infancy.

Both approaches agree that voluntary grasping starts to occur around 3- or 4-months of age and undergoes major transformations between 4- and 13-months. During this time, we learn to grasp and hold objects using basic handshapes that we will then continue to use and refine in different contexts throughout our lifetime (for a classification of adult handshapes in grasping see Cutkosky 1989). However, handshapes characterizing grasping do not all appear at the same time. They undergo significant changes and are subject to a gradual selection process that is largely driven by environmental factors. These changes are important because they allow us to better understand not only the primacy of specific handshapes over others (something that we will see reflected both in gestures and signs), but they also influence our relationship with objects and learning object functions, which is an essential building block towards the emergence of symbols as we will see below.

Another aspect on which the different approaches converge is that infants start grasping objects by using the palm of the hand (palmar grasp) and that initially, this behaviour is irrespective of object size or shape. The advantages of palmar grasps are evident: the palm offers a broader contact surface compared to individual digits, therefore requiring less accuracy in planning the reaching phase; placing the whole hand on the object and closing it around it is generally sufficient. Palmar grasps are also very stable. They have been called "power grips" because the object may be held firmly avoiding slips and maintaining a stable position in the hand (Napier 1956, 1980). Between 3- and 4-months of age, infants explore different types of palmar grips and, similarly to what we have seen in reflexes, initial changes in voluntary grasping involve "shifting" of the object within the palm. Objects are initially grasped with the ulnar part of the hand at 3-months, only to shift radially around 6-months and then distally around 9-months (Connolly and Elliott 1972). The first year of life, therefore, is characterized by significant "shifts" in the areas of the hand involved initially in grasping reflexes and subsequently in voluntary grasping, which is characterized by an ulnar-radial shift and a proximal-distal shift, as shown in Fig. 2. As we shall see in the following paragraphs, these changes will prove important in the emergence of symbols.

As simple and stable as the palmar grasp may be, the handshapes that it allows have various limitations. As the entire hand surrounds the object, this leads to poor visibility of object parts, limited use of in-hand mobility and reduced freedom in adjusting the grip to exploit object parts in dedicated actions. This means that if we are using a palmar grasp and we wish to explore an object while holding it or shift its position within our hand or grasp a specific part to use it within an action, we

**Fig. 3** Examples of palmar versus precision grips for a small cube and a small bell (Adapted from Halverson 1943)

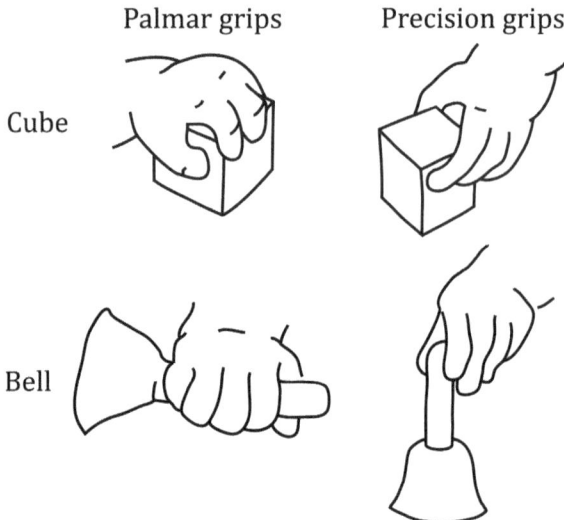

will need to use our other hand and make some bimanual adjustments. A behavior that is often observed early in development is an unspecific placement of the hand on an object, the production of a palmar grasp that allows object retrieval (which covers most of the object's surface) followed by bimanual object exploration. However, using two hands to perform one task is hardly an efficient behavior. For one thing, it means that we can only deal with one object at the time. To move beyond the limits of palmar grips, infants need to make what may be considered one of the major steps in evolutionary history: they need to produce precision grips.

Precision grips do not rely on the palm but on the digits. Given the anatomy of the human hand, precision grips allow for a multitude of hand configurations (Cutkosky 1989). They also support in-hand adjustments and manipulation, proving more effective towards fine-grained object exploration and use (Napier 1956) (see Fig. 3). According to Halverson's analysis, infants between 4- and 13-months of age show significant progress in the production of precision grips. Objects initially start to "shift" radially within the palm around 7-months, and thumb opposition starts to emerge. A subsequent distal "shift" allows other digits to be enrolled in grasping, leading to the emergence of precision grasps between 9- and 13-months (Halverson 1943). Precision grips are refined throughout our lifetime and encounters with new tasks, and objects play an important role in this process.

Halverson's linear description of the gradual emergence of precision grips ignores; however, the co-occurring development of other cognitive skills and the active role of objects, tools and task spaces. With the exception of the famous character called "Thing" in the Addams Family television series, hands typically do not walk around on their own but are rather part of a body that is situated within the environment. In the case of grasping, both the body and the environment matter for the emergence of gradually more complex hand configurations. As noted by Iverson

(2010) motor milestones (e.g. unsupported sitting, reaching, crawling and later walking), extend the infant's world beyond here and now, allowing encounters with new and diverse objects, tools and social contexts, all of which contribute to shaping and refining grasping. Therefore, the conquest of fine-tuned precision grasps cannot be considered in isolation, but must take into consideration the development of other motor and cognitive skills—that is, not only linked to locomotion but also to visual guidance, reaching, motor planning, proprioception, etc., as well as to the active role of objects, tools and task spaces in building possibilities for action. Objects and tools not only offer different affordances,[2] scaffolding the evolution of gradually more complex hand configurations (e.g. think of grasping a spoon vs. grasping a cube) but are organized within specific contexts (e.g. feeding, dressing, etc.), which structure the functional use of tools and actively shape infants' environments supporting different types of object exploration and use (Gibson 1979; Kisch 1996).

Relevant here is the study by Newell and colleagues (1990), which analysed hand configurations in a cross-sectional study of five age groups (i.e. 4-, 5-, 6-, 7- and 8-month olds) with a total of 107 infants. The task required grasping objects that varied in size, shape and positions (e.g. a wooden cube or a lightweight toy cup, which could be presented facing upwards or downwards). Results showed that no relevant differences emerged among 4- to 8-month olds in types of grip configurations used. Grip configurations did change, however, as a function of object properties across all age groups tested, with 4-month olds being more variable in the type of grips used. This finding goes against Halverson's strictly maturational view of grasping and shows that greater changes in handshapes may be seen in relation to object properties and orientation than as a function of time.

In their beautifully fine-grained analysis of spoon-feeding in infants, Connolly and Dagleish (1989) highlight the importance of grasping technique, tools and functional actions in handshape formation. The authors show that infants between 11- and 23-months produce a total of 11 different grasp types, which tended to diminish in variety over time—i.e. younger children producing a greater variety of grasp types, with respect to older ones. This result provides further support to the argument that object and task constraints play an active role. Older infants, who could potentially produce a huge number of grasp configurations, actually produce fewer grasp types within given contexts. Further, analyses of grasp types produced by older infants showed that handshapes that make spoon use more difficult were gradually abandoned, while those facilitating feeding actions were maintained. Therefore, while infants were progressively reducing the number of different grasp types, they were also converging on those types that were more effective in carrying out the object's main function of feeding (Connolly and Dagleish 1989). This is a very relevant point as it illustrates how learning to grasp an object using the most

---

[2]The concept of "affordance" was initially introduced by Gibson (1979) to indicate invariant features in the environment that offer themselves to an organism supporting or "affording" perceptual or sensory-motor exploitation. For example, a mug may have multiple affordances with a handle that supports grasping and a hollow part that allows carrying liquids.

effective handshape goes beyond the mere physical characteristics of the object and its affordances, projecting the infant into the world of functional actions with tools. Getting to know objects, their affordances and functions is an essential stepping stone towards symbol formation, as we will see in the following discussion.

So far, we have only considered typical development, but handshapes are also important in studies of atypical development. They are present in many standardized measures of infant skills (e.g. the Muller Scales for Early Learning, MSEL, the Motor Assessment Battery for Children, M-ABC), which allow detection of cases of atypical development such as apraxia or motor delays in specific populations. For example, infants at high risk for autism spectrum disorder (ASD) (i.e. infants with an older sibling with ASD) showed reduced overall grasping skills as measured by the MSEL (Libertus et al. 2014). Furthermore, handshapes may also tell us more about communicative development. Recent studies have highlighted the link between motor skills and language development. For example, showing that atypical gross motor skills (e.g. independent sitting, rhythmic arm movements, postural control), may negatively impact communication and language development (Bhat et al. 2012; Iverson 2010). If grasping and handshapes support the emergence of symbols and communicative acts (as we try to show in the discussion below), evidence of delays or impairments in these skills within atypical populations characterized by communication deficits (e.g. autism spectrum disorders, Williams syndrome) bring whole new implications.

In this regard, it is interesting to consider a study on infants at high risk for autism spectrum disorders (ASD). These are infants who have an older sibling with ASD and are at heightened biological risk for the disorder (i.e. they are 18.7 times more likely to develop ASD than the general population) and may or may not receive an ASD diagnosis at 36-months; therefore, they allow to study the presence or the absence of specific early emerging delays prior to age of diagnosis (Ozonoff et al. 2011; Rogers 2009). Sparaci and colleagues (forthcoming) analysed the emergence different grasp types during spoon use in a longitudinal study of 41 infants at high risk for ASD between 10- and 36-months. Interestingly, infants that received an ASD diagnosis at 36-months were unable to produce functional actions at 10-months and produced fewer grasps facilitating tool use at 24-months compared to infants who showed no delay at 36-months. Furthermore, functional action production at 10-months was a strong predictor of words produced at 24-months even when controlling for cognitive level.

A study on Williams syndrome (WS) by Stefanini and colleagues (2008) used the M-ABC to assess motor skills in children with WS between 9- and 11-years of age. Their results highlighted differences in hand configurations during the execution of precision grips in children with WS: they produced a greater number of lateral grips (i.e. using the pad of the thumb placed on the outside of the index finger) rather than precision grips (i.e. the pad of the thumb placed on the tip of the index finger) than both chronological- and age-matched peers (Stefanini et al. 2008).

Recent studies considering older children with ASD and children with WS, show that difficulties in the execution of grasping are mirrored by corresponding difficulties in understanding others' grasping acts in terms of their functional meaning (i.e. understanding *why* an object is being grasped) (Boria et al. 2009; Sparaci et al. 2012; Sparaci et al. 2014). This may have important effects on children's ability to understand others actions, with a negative impact on social interactions and communication development.

Summing up the discussion thus far, we have described the emergence of hand configurations in grasping during infancy, showing how their emergence is strongly influenced by organismic, environmental and task constraints. We have begun to delineate how learning to grasp objects establishes a link between concrete actions and more abstract concepts, such as object functions. We have also mentioned how different patterns of hand configurations may be observed in atypical development. We will now move on to consider the role of hand configurations in communication and in the emergence of symbols.

## 3 Handshapes in Gesture Research: The Emergence of Intentional Communication and Symbols

How can handshapes and grasping objects play a role in something so abstract as the emergence of symbols? While this is a very broad question that still calls for further investigation, some initial answers were offered in the mid-seventies when some researchers challenged widely accepted Chomskian approaches by taking a multimodal approach to language. In relation to development, these researchers proposed that in order to fully understand language we must move beyond grammar and begin "well *before* language begins" concentrating on the sensory, motor, conceptual and social prerequisites that make language possible (Bruner 1975, p. 257). Jerome Bruner, strongly influenced by Speech Act Theory,[3] proposed that children's ability to perform and attend to motor actions may lead them to grasp concepts that are later embodied in language (Bruner 1974). In the same years, Bates et al. (1975), began to investigate prerequisites to communication in a longitudinal study of three infant girls who were at the beginning of the study 2-, 6- and 12-months old and were observed in home visits at two week intervals over a period of eight months. This study allowed distinguishing three main "pre-speech" phases in development. In the first *perlocutionary stage* (from age 2- to 10-months),

---

[3]Speech Acts Theory was initially formulated by the philosopher John Austin in a famous series of lectures collected in *How to Do things with Words* (Austin 1962). Austin showed how, in the pragmatics of communication, sentences can be analysed not only in terms of information conveyed, but as bringing forth three types of actions: the act of saying something (i.e. the *locutionary act*); what one wishes to communicate *in* saying it (i.e. the *illocutionary act*); what causal consequences one wishes to bring about *by* saying it (i.e. the *perlocutionary act*). This distinction was later taken up and extended by John Searle who suggested that every statement can be analysed according to its propositional content (i.e. *locutionary act*) and performative content (i.e. *illocutionary act*) (Searle 1969; see also Sparaci 2010 for a short review of the relation between Austin's theory and Bruner's work).

intentional communication is still absent, but infants start exploring objects by handling and mouthing them. Caregivers were shown to interpret these object explorations and internal states (e.g. smiling, crying) as early forms of means–end relationships. The second *illocutionary stage* (from age 10- to 13-months) saw the emergence of gradually more complex "performative structures". Within these structures, "actions that were originally means for reaching the goal itself (e.g. orienting, reaching, grasping) were gradually separated from the concrete attempt to reach objects, and became instead signals" (Bates et al. 1975, p. 219). In the progression towards gradually more complex performative structures, the authors traced the gradual emergence of intentional communication (see Table 1). Finally, during the third *locutionary stage*, word-like signals with corresponding referents start to emerge out of the same action schemes characterizing the illocutionary stage.

Looking at the behaviors described by Bates and colleagues as characterizing the emergence of performatives (see Table 1), we can begin to understand how early action sequences with objects, involving different types of handshapes (e.g. requesting with an open hand with palm facing upward, showing an object on the open palm, pointing with the whole hand or the index finger, etc.) soon become early forms of communication. For example, requesting, showing, giving and pointing (also termed deictic gestures) are strongly dependent on infants' having acquired the necessary motor skills appropriate to specific communicative contexts. At this stage, infants are merely expressing intent to communicate, while content is provided by contextual cues (e.g. the presence of a specific object in a specific situation). A good example of this link between motor skills and early communication may be found by considering the emergence of pointing.

Consistent use of full-blown index pointing[4] emerges at around 11- or 12-months, often accompanied by vocalizations, but this skill is preceded by previous motor patterns that emerge very early in development. Antecedents may be observed in infants before the first year of life in which hand movements and early vocalizations are highly related (Carpenter et al. 1998; Leung and Rheingold 1981, Behne et al. 2012; Bernardis et al. 2008; Gentilucci et al. 2008; Iverson and Thelen 1999). For example, early index finger extensions without an outstretched arm movement can be observed during mother–infant interactions at 2-months of age (Fogel and Hannan 1985; Masataka 2003), and pointing to explore proximal objects by poking is present at 9-months (Bates et al. 1979). After 10-months, infants are able to point towards proximal objects, while pointing at distal objects emerges only later, around 13-months, indicating that development of appropriate patterns of visuo-motor coordination is essential in acquiring this skill (Butterworth 2003; Franco and Butterworth 1996). This is also evident in infants' understanding of pointing, which precedes its production and which begins to appear already around

---

[4]Full-blown pointing is usually defined as a gesture in which "the index finger and arm are extended in the direction of the interesting object, whereas the remaining fingers are curled under the hand, with the thumb held down and to the side" (Butterworth 2003, p. 9).

Hands Shaping Communication: From Gestures to Signs

**Table 1** Summary of the main performative structures that emerge during the illocutionary stage with examples of child behaviour, as described in Bates et al. 1975

| Performative | Description | Example | Age at observation (months) |
|---|---|---|---|
| Showing off | The infant repeats behaviors that have been previously successful in provoking laughter in the caregiver | "[Carlotta] is in her mother's arms, drinking milk from a glass. When she has finished drinking, she looks around at the adults watching her and makes a comical noise with her mouth [...] adults laugh and [Carlotta] repeats the activity several times" (Bates et al., 1975, p. 216) | 9 |
| Showing | The infant presents objects that are within her/his grasp on an open palm, without the intention of giving them to the adult | "[Carlotta] is playing with a toy already in her hand; suddenly, she looks towards the observer and extends her arm forward holding the toy [...] when the adult tries to take the exhibited toy, [Carlotta] refuses to let go, and often pulls her arm back" (Bates et al., 1975, p. 216) | 10 |
| Pointing for self | The infant uses pointing while alone in a room and unaware of others | "[Carlotta uses pointing] in examination of small book figures, and in orienting towards novel and interesting sounds and events. However, these earlier pointing sequences never involve a search for adult attention. Carlotta may use them while alone in a room and unaware of our observation" (Bates et al., 1975, p. 217) | 12 |
| Give | The infant gives an object to the caregiver (separate from showing) | "[Carlotta] takes a wooden mask from a chair, crosses the room smiling and looking at the observer, and drops the mask in the observer's lap." (Bates et al., 1975, p. 217) | 13 |
| Pointing for others | The infant performs a sequence of actions which include pointing and involve seeking the caregiver's attention | "[Carlotta] would first orient towards the interesting object or event, extending her arm forward and forefinger in the characteristic pointing gesture while uttering a breathy sound "ha". Then she would swing around, point at the adult with the same gesture, and return to look at the object and point towards it once again" (Bates et al., 1975, p. 217) | 12–13 |

9-months, although it is strongly influenced by target distance and movement, becoming fully developed only between 12- and 18-months (Grover 1988; Morrisette et al. 1995; Behne et al. 2012).

Deictic gestures are, however, strongly linked to the context in which they are performed, whereas symbols have the property of maintaining their meaning in the absence of the referent. This comes back to our basic question of how can something as abstract as a symbolic form emerge from a concrete action? Relevant here are various studies that have observed that at the same age at which infants' ability to manipulate objects reaches a very important developmental milestone with the appearance of precision grips (around 12-months), and just before spoken naming onset, children start to reproduce short action sequences with objects, usually related to the object's function, for example, the child could place a comb or brush on her/his hair (Bates et al. 1979; Caselli 1990). Soon afterwards, these action sequences begin to be performed in the absence of objects, while maintaining their meaning (e.g. the child could place his/her empty hand on his/her hair), as confirmed by studies using the McArthur–Bates CDI (Caselli et al. 2012).[5] As already discussed above, during infancy, grasping objects and tools allows the emergence of gradually more complex hand configurations, and these configurations are strongly linked to functional actions with objects while being influenced by organismic, environmental and task constraints. As the infant's proficiency with tools grows, she/he is able to interact more effectively with objects and to explore different contexts for action (Iverson 2010). This also means that specific grasps and hand configurations may be used within a growing number of situations and thus gradually become detached from the specific contexts in which they were initially acquired. For example, infants have been shown to be more effective in grasping a spoon in order to feed themselves and only later in order to feed others (McCarty et al. 2001). Similarly, actions initially carried out within a real-life context (i.e. grasping a spoon to feed oneself or another) are often carried over into pretend play situations (i.e. feeding a doll with a spoon). Bates and colleagues (1979) underscore how performing actions with tools outside their original contexts of acquisition supports the emergence of symbols in a progression from "presentational" to "representational" symbolizing which is present both in language and in actions. Throughout this process, the environment plays an active role in shaping actions, as described by Bates and colleagues:

> The child takes an increasingly active role while the amount of perceptual support in the environment becomes increasingly important. For example, in symbolic play the child may stir with a real spoon in a bowl, in the kitchen where Mother is doing similar spoon things. This would be an example of presentational symbolic activity. In representational

---

[5]The McArthur–Bates Communicative Development Inventories (CDI) are parent questionnaires assessing action and gesture production, word comprehension, word production and early grammatical development between 8- and 36-months (Fenson et al. 2007). They are valid and reliable measures that have been adapted to a number of languages and cultures other than English and used across the world in children with both typical and atypical development. Further information may be fond on the following website: http://mb-cdi.stanford.edu/index.html.

symbolizing [...] the child may stir with an empty hand, using a block for a bowl, while playing in the bedroom with no adults around. This developmental progression parallels developments in tool use [...] (Bates et al. 1979, p. 326)

Lifter and Bloom (1989) also showed how object displacement activities in which infants use one object in relation to another (e.g. feeding a doll with a spoon) are closely linked to achievements in language.

Therefore, it seems clear that motor repertoires and handshapes exploited by infants in early grasping appear to be the basis not only of highly context-related deictic gestures (e.g. pointing) but also of representational gestures that are performed in the absence of an object and denote a specific referent while remaining relatively stable across different contexts (Capirci et al. 2005). Representational gestures are very important during early phases of communicative development. Their use across cultures has been confirmed in various studies. (e.g. see Zinober and Martlew 1985 for British children, Acredolo and Goodwyn 1988, for American children and Pettenati et al. 2012 for Italian and Japanese children). They account for a large portion of children's communicative repertoires and support the emergence of two-element utterances. For example, between 16- and 20-months of age, two-word utterances are mostly made up of gesture–word combinations (Goldin-Meadow and Morford 1990; Capirci et al. 1996; Butcher and Goldin-Meadow 2000). In a small-scale, longitudinal study of three children between ages 10- and 23-months, who were videotaped monthly during spontaneous play sessions with a caregiver, Capirci and colleagues (2005) showed continuity between early action schemes, gestures and words. Meaningful communicative actions were produced by the children in this study from the first session in which representational gestures were used and then gradually increased with age, whereas single word and gesture–word combinations first appeared between 12- and 16-months. Also relevant here is the authors' finding that meaning correspondence between early action patterns (e.g. such as a pushing motion for pushing a little car), representational gestures and later utterances (e.g. "brum brum"), indicate "that the emergence of a particular action preceded the production of the gesture and/or word with the corresponding meaning" (Capirci et al. 2005, p. 171). Even if this study involved only a small sample of children, it does point towards a strong link between early actions with objects and later symbolic forms.

Another way of assessing this link is to consider the role of motor constraints. We have seen that during early phases of development, children are still fine-tuning their ability to produce specific handshapes and that this attuning depends both on motor development and on contexts of action. If handshapes produced in grasping objects are the basis of handshapes produced in gestures, then we should find that these gestures are also affected by similar motor constraints on the handshapes. Pettenati and colleagues investigated the role of motoric constraints on the production of representational gestures in a picture-naming task done with 45 children between 2- and 3-years of age (Pettenati et al. 2009). In this study, gestures were analysed using the formational parameters commonly used in analyses of early signs (of which more will be said later). This is a helpful technique in gesture studies and has also been used in adult studies, for example, in analysing gestures made by classical orchestra conductors (Boyes Braem and Braem 2000). Pettenati and colleagues (2009) showed that children's gestures relied on a restricted set of "basic" handshapes. Among these, the most common ones either resembled a palmar grasp (e.g. closed fist), involved an open palm (e.g. fingers open with flat hand) or pointing (i.e. all fingers closed and index finger extended, see handshapes A, 5 and G in Table 2). Handshapes resembling precision grips (e.g. thumb touching index finger and all other fingers closed) were less frequently used (see handshape O in Table 2). Obviously, handshapes produced in this study may also have been influenced by contextual factors, for example, the specific pictures shown to children or the spoken words accompanying the gesture. However, it is interesting to note not only that these young children resorted to basic handshapes but also that in the majority of two-handed gestures, the hands were symmetrical, confirming the observation that motorically complex gestures requiring separate control of the two hands emerge at a later stage (Pettenati et al. 2009).

Pettenati and colleagues, by using in their analysis of hearing children's gestures the same parameters commonly used to analyse deaf children's signs, showed that hearing children's co-speech gestures are similar in function and handshape to deaf children's early signs. This result strengthens the hypothesis of a link between motor skills and linguistic systems at this early developmental stage. Furthermore, children (and adults) who are already able to produce the verbal label for a given object or action continue to use gestural enactment or depiction to accompany their naming acts, thereby representing the object or the action and indicating functional similarities between representational gestures and lexical signs. These findings support the view that motoric characteristics of handshapes operate both in the production of signs and in the production of gestures and could be largely explained by the anatomy and physiology of the hand and arm (Ann 1996).

Before looking more closely at handshapes in sign languages, it is important to observe how links between handshapes used for grasping and in gestures go beyond data description to encompass similarities in theoretical approaches. In the studies on grasping development described above, we have seen how authors using a maturational approach tend to describe handshapes as emerging in a linear progression while choosing to overlook the role of objects, tools and contexts. Something similar has happened in studies analysing handshapes in representational gestures focusing on representational techniques.

Representational techniques describe the way in which a representational gesture "depicts" an object or action.[6] Two representational techniques described in relation to representational gestures focus on handshape as a key issue. The first one is termed "hand-as-object" in order to indicate cases in which the child's hand represents the object itself or its shape (e.g. an open hand with parted fingers "brushing" through the hair may be used to represent the shape of a comb and its

---

[6]For a full description of representational techniques see also Marentette et al. 2016.

teeth in the gesture COMB[7]). The second is termed "hand-as-hand", to highlight cases in which the child's hand depicts the way in which a given object is held or used (e.g. a closed fist with the palm downwards passed on the head in a "brushing" motion may be used to represent the way in which a comb is held in the gesture COMB). Since hand-as-object gestures represent salient aspects of an object's shape, they have been considered less symbolic in research on gestures than hand-as-hand gestures, representing objects' functions, but not their shape. Researchers have, therefore, tended to stress a developmental trend in which hand-as-object gestures precede hand-as-hand gestures in both production (Boyatzis and Watson 1993) and comprehension (O'Reilly 1995). Findings from elicited pantomime studies showing that when asked to pretend to use a comb children under 6-years of age tended to produce hand-as-object gestures, while children older than 6 produced more hand-as-hand gestures, were taken as evidence of a limit in pre-schoolers' symbolic capacities and a shift from form-based to function-based handshapes and representations (Overton and Jackson 1973; O'Reilly 1995; Dick et al. 2005).

However, recent studies have started to challenge this "maturational" approach to handshapes in representational gestures, questioning the existence of a linear progression and highlighting instead the importance of context. In this approach, it appears that chosen handshapes may depend rather on the object or action that is being represented, on the way in which the gesture is elicited and/or on the communicative contexts in which the gesture is being used. A recent study by Marentette and colleagues examined spontaneous gestures produced by 22 Italian children and 42 Canadian children in a naming task and found that two-year olds were equally likely to produce hand-as-hand gestures as hand-as-object gestures. Differences in representational technique were related not so much to age but rather to the context in which the gesture was produced (Marentette et al. 2016). For example, while almost all gestures were produced in conjunction with speech and in association with predicate rather than nouns target pictures; hand-as-object gestures were more likely to be associated with the production of nouns and to small objects (e.g. UMBRELLA, COMB, FORK), while hand-as-hand gestures were usually used in relation to pictures representing an agent performing an action with a large object (e.g. DRIVING, OPENING, PUSHING) (Marentette et al. 2016).

Hence, there seems to be not only a link with grasping handshapes and early motor skills but also a strong influence of context, both linguistic and environmental on representation techniques and gesture types. Between 4- and 5-years of age, children use idiosyncratic, content-loaded gestures in narratives and conversations, and closer analyses have shown interesting differences in gesture types (Capirci et al. 2011). In later childhood, other types of gestures emerge (i.e. "rhythmic" or "emphatic" gestures) (Gullberg et al. 2008). For example, younger children produce very few metaphoric, abstract deictic gestures and beats, which are

---

[7]In the present chapter, all gestures and signs will be reported in capital letters extending to gestures a convention usually employed in sign language studies for signs.

seen more frequently in the production of older children (Colletta 2004). Similarly, studies on problem-solving tasks (i.e. reasoning about balance or mathematical equivalence) show that school-age children convey a substantial proportion of their knowledge through co-speech gestures and that gesture–speech "mismatches" predict their learning and memory skills, a phenomenon also reported in adulthood (Alibali and Goldin-Meadow 1993; Church and Goldin-Meadow 1986; Pine et al. 2004; Goldin-Meadow 2017; Gurney et al. 2013). These studies show that gesture use changes according to contexts of use as well as according to developmental stages and that any study on the emergence of symbols must consider both aspects in order to fully understand the richness of multimodal and embodied communication.

In tracing a continuum from gestures to full-blown symbols, Adam Kendon described a progression in gestures types that were later termed by David McNeill as *"Kendon's continuum*: Gesticulation → Language-like Gestures → Pantomimes → Emblems → Sign Languages" (McNeill 1992, p. 37). At the far end of the continuum, we encounter conventionalized signs in sign languages, which will be the topic of the next section.

## 4 Handshapes in Sign Language Linguistics

Since antiquity, deaf people who have difficulties communicating via the acoustic-spoken modality have developed languages that exploit the possibilities of the visual–gestural modality The Ethnologue database (https://www.ethnologue.com/subgroups/deaf-sign-language) in its latest edition lists 138 sign languages (SLs) used in various parts of the world. These SLs are widely recognized by the scientific community as full-fledged, natural languages and include Italian Sign Language or LIS (Volterra 1987; Pizzuto and Corazza 1996; Geraci 2012). It has been shown that, even though these languages are perceived and produced in the visual–gestural (rather than in the vocal auditory) modality, they satisfy the communicative and expressive needs of a community and possess all basic linguistic components, including a phonology, lexicon, syntax and grammar. All of these SLs involve the linguistic use of manual and non-manual components (i.e. face, eye gaze, posture and body movements). Here, we will focus mainly on the use of the hands in SLs.

In the flow of information conveyed through a SL, it is possible to recognize specific signs, which are considered as conventionalized lexical units and are listed in SL dictionaries (see also Radutzky 1992 for Italian Sign Language—LIS; Sutton-Spence and Woll 1999 for British Sign Language—BSL). Each lexical sign denotes a specific meaning, much as does a word of a spoken language. Since the pioneering work of William Stokoe on American Sign Language (ASL), linguistic research has shown that the structure of each sign unit is analysable into subcomponents (Stokoe 1960). All signs in every SL are formed by combining a defined number of formational parameters, just as words of a spoken language are formed on the basis of phonemes in various combinations. Each manual sign can be broken down into four basic parameters: the hand configuration or handshape; the orientation of the hand; the location in which the sign is performed; and the movement of the hands. Within these parameters, there are limited sets of subcomponents used in individual SLs.

What is special about this visual language is that many of these simultaneously produced subcomponents analogous to phonemes can also carry separate meanings —i.e. the same form can in some cases function purely "phonemically" to distinguish between forms but in other cases also carry meaning as do morphemes. For example, in some "polymorphemic" signs such as TO-GIVE, the movement component can convey a meaning of "from whom/to whom" while the handshape component can refer to *what* is being given. For example, a horizontal movement of the hand in a closed fist hand configuration (see handshape A in Table 2) could be used in referring to giving an object like a mug held by its handle, whereas the same movement performed with an open semicircle handshape (see handshape C in Table 2) may be used to represent giving a glass held by its side. The beginning and end location components may be used instead to specify *who* is giving to whom. As pointed out by, among others, Boyes Braem (1981), many conventional lexical units are made up of formal features that are visually motivated and thereby iconic. Their visual motivation is not random or idiosyncratic but derives from regularities at the level of formational parameters. Handshapes, for example, are often linked to features of a sign's meaning via reference to some peculiar visual forms (for more examples in LIS see Pizzuto et al. 1995; Pietrandrea and Russo 2007). The handshapes themselves are not tied to a specific meaning but are polysemous in that they are capable of conveying several meanings, depending on the context of the other parameters. Of the many handshapes that humans are physically capable of performing, only a limited number is used linguistically in SLs. See Table 2 for some examples.[8]

Some handshapes are particularly interesting as they are examples of how some hand configurations may be linked to a particular culture. For example, the "horn" handshape has a negative meaning in Italian culture, which is reflected in the fact that it appears in LIS signs such as "devil", "temptation" and "spite" (see Table 2). The same does not hold true in other cultures. The I handshape (see Table 2) has a negative meaning in British culture, but not in Italian culture. In BSL, this handshape is often used in signs with a negative meaning such as "swear", "awful", "criticize" while in LIS it represents often a thin object, and it is used in signs such as "spaghetti", "skinny" and "string". A study on how several LIS signs with transparent vs. non-transparent meanings were understood by both hearing non-signers and SL users from six different European countries (Boyes Braem et al. 2002) has shown how these cultural differences in handshapes can cause sign/gesture interference and misinterpretations. For example, for the LIS sign meaning "to go out" (a sign executed in the neutral space with the two hands with the same handshape, see

---

[8]All handshapes presented in this paper are represented using the annotation system described in Eccarius and Brentari 2008.

**Table 2** Examples of some LIS handshapes and of visual metaphors underlying them (handshape labels refer to LIS)

| Handshape | | Sample visual forms/metaphors | Corresponding sample signs in LIS |
|---|---|---|---|
| | A | power or strength | "strength", "courage" |
| | | grasping or handling | "bicycle", "bag" |
| | | a solid and/or compact round object | "stone", "football" |
| | B | personal, spatial and temporal deixis | "myself", "here", "now" |
| | | level, straight, extended, non-permeable surfaces | "door", "table", "wall", "book" |
| | | borders of wider objects or areas | "frame", "room" |
| | | the action of cutting a real object or an abstract concept | "slicing", "dividing" |
| | F | very defined real or virtual dot-like objects | "freckles", "objective" |
| | | small and light objects which are usually handled using a pincer grip | "flower", "sheet", "sewing" |
| | | entities which may sting in a real or figurative manner | "mosquito", "critique" |
| | T# | objects with a graspable stem or rod | "artichoke", "icecream", "fishing" |
| | 5 | transparent objects, permeable surfaces | "water", "mirror" |
| | G | residual form of pointing | "cry", "deaf" |
| | | long and thin objects or animals | "snake", "worm" |
| | | entities which are penetrating in a figurative manner | "jealousy", "envy" |
| | C | circular and graspable objects | "glass", "bottle", "tourism" |
| | O | small round objects or objects with a small round section | "binoculars", "buttons", "courgette" |
| | ⊔ | animals with horns or horn-like shape | "snail", "bull" |
| | Y | entities with two protruding parts or their derivatives | "cow", "plane", "to fly by plane" |
| | 3 | entities with three or more protruding parts | "frog", "jump", "goose", "rooster" |
| | V | entities with two or more protruding parts | "scissors", "salad", "to cut", "scissors", "fork, plug" |
| | 4 | sequences of thin or small entities or parts of entities | "king", "mathematics", "procession" |
| | H | entities with an overall small and linear appearance | "fettuccine", "fish", "name", "butter" |
| | I | thin objects or people | "slim", "spaghetti" |
| | | entities which are sharp in a concrete or figurative manner | "pain", "kill" |

For a full description of handshapes which are recognized as distinctively linguistic in LIS see Volterra (1987)

For a full description of handshapes which are recognized as distinctively linguistic in LIS see also Volterra (1987) handshape Y in Table 2, moving outward), the hearing participants ignored the handshape but paid attention to the direction of the linear movement (a slight curve moving out in front of the signer) and their answers mainly referred to the first persons (i.e. "you", "everybody", "yours"). In contrast, British deaf participants, who, being signers, pay much more attention to the handshape, which in this case has a conventionalized meaning in their SL, interpreted the LIS sign with negative meanings such as "regret", "terrible", "dead", "awful" or "bad".

The handshapes found in SLs are not only linked to specific meanings in different cultures but also seem to emerge at different stages of acquisition. On this point, a very interesting early study by Boyes Braem (1994) provided a stage model describing the acquisition of handshapes in ASL in a child aged 2-years and 7-months (see Table 3). Handshapes belonging to the first stage are all well known to the child before she/he starts using them as signs, as they are all motorically related to handshapes that we have discussed above as being produced by the pre-linguistic infant in reaching (e.g. handshape 5 in Table 3), palmar grasps (e.g. handshape C in Table 3), probing or pointing (e.g. handshape G in Table 3) and basic precision grips (e.g. handshape bO in Table 3). In sign language linguistics, these are all commonly referred to as "unmarked" handshapes as they involve few selected features and are

**Table 3** Examples of ASL handshapes during the first, second, third and fourth stage of acquisition as described in Boyes Braem (1994) (handshape labels refer to ASL)

| Handshape types | Stages of acquisition | Handshapes | | | |
|---|---|---|---|---|---|
| Unmarked | Stage 1 | A | As | L | bO |
| | | G | 5 | C | |
| | Stage 2 | B | F | O | |
| Marked | Stage 3 | I | Y | D | P |
| | | 3 | V | H | W |
| | Stage 4 | Open 8 | 7 | X | |
| | | R | T | | |

motorically less complex. Handshapes characterizing stage 2 are motorically more complex, involving either a rigid extension of all fingers (e.g. handshape B in Table 3) or more complex forms of precision grips (e.g. handshape F in Table 3).

Stages 3 and 4, in contrast, contain so-called marked handshapes, which involve a greater number of selected features and are motorically more complex, requiring inhibition and extension of the middle, ring and pinkie fingers, as well as control of non-adjacent fingers. Motoric constraints are not the only factors, which play a role in the emergence of signs. Frequency of a given handshape in a SL may also be important as children may be more exposed to specific frequently occurring handshapes. However, there also seems to be a link between how easy it is to produce a given handshape and its frequency. Unmarked handshapes are the ones appearing most frequently in all of the lexicons of the world's different SLs (see also Sutton-Spence and Woll 1999 for BSL; Pietrandrea 1997, for LIS), whereas marked handshapes have a more varied distribution. Later studies have provided supporting evidence for the primacy of a subset of handshapes present in Boyes Braem's first two stages of development, not only for children acquiring ASL (Conlin et al. 2000) but also for other SLs (among others, Karnopp 2002; Morgan et al. 2007; McIntire 1977; Carter 1981; Von Tetzchner 1984).

These findings, taken together with those on motoric characteristics of gestures described above, point to the importance of considering motor constraints alongside other factors in the analysis of handshapes in SLs. The relationship to gestures, however, extends beyond this to include the distinction introduced above between representational techniques: hand-as-hand and hand-as-object. Brentari and colleagues (2014) analysed contrasts between handling classifiers (i.e. "hand-as-hand" handshapes that represent how objects are manipulated) and object classifiers (i.e. "hand-as-object" handshapes that represent the class, size or shape of objects) comparing productions of signers (in ASL and LIS) and gesturers using silent gestures (in Italian and American). Previous studies on SLs had shown that handling handshapes and object handshapes express an agentive/non-agentive semantic distinction in many SLs, with handling handshapes used in agentive event descriptions and object handshapes used in non-agentive event descriptions. Brentari and colleagues confirmed the agentive/non-agentive handshape for signers of ASL and LIS, but also observed how some individual hearing gesturers produced the same opposition for agentive and non-agentive event descriptions. This was more the case for Italian than American adult gesturers. The authors conclude that cognitive, cultural and linguistic factors contribute to the conventionalization of this distinction of handshape types.

As mentioned above, Marentette and colleagues (2016) also showed that the same basic representational strategies observed in studies of adult communicative gesture and children symbol use and in sign research are already present in the representational gesture repertoires of hearing two-year olds from two cultural groups: Italian children (from a high-gesture culture) and Canadian children (from a low-gesture culture). The use of representational techniques to make visible different types of embodied practices suggests that organizational techniques for depicting information about objects and events have a shared cognitive basis that is recruited by both language and gestural systems.

## 5 Summary and Conclusions

In showing how children slowly progress from reflexes to object grasping, we have laid out a path from early motor actions to the emergence of differentiated handshapes. We then proceeded to highlight how these handshapes play an important role in symbol formation both in gestures and in signs. In particular, we stressed the importance of considering factors beyond motoric development to include contextual specificity (in both grasping, gesture and signs). Our main aim here has been to provide stronger grounding for an embodied and multimodal approach to language, which would consider not only speech but also the relevant role of the hand and its *shapes* in human communication.

Two notions have often been used in archaeology and human evolution studies as relevant for understanding human nature: the notion of *Homo faber* (the "practical maker") and *Homo symbolicus* (the "symbol maker") (Malafouris 2013). *Homo faber* refers to the fact that despite a great number of studies on nonhuman primates and their abilities to select and use tools in causally complex ways, human tool production and use surpasses in variety and complexity that of all other species (Visalberghi and Tomasello 1989). The notion of *Homo symbolicus* highlights the fact that we are essentially a "symbolic species" (Deacon 1997). In their 1979 book *The Emergence of Symbols*, Bates and colleagues brought these two notions together in a unique way in their focus on the emergence of communication in infancy. In this chapter, we have attempted to follow the path laid out in this earlier work and to enrich it by bringing in recent literature that gives more documentation for the link between motor actions, gestures and signs, here with a particular focus on handshapes (see also Volterra, Capirci, Caselli, Rinaldi and Sparaci 2017 for a recent review of research on the action, gesture, sign continuity).

An interesting question here is why has the continuity of handshape from grasping, to gestures and to signs not been specifically highlighted before. One reason could be the historical distinctions between different fields of research. Only recently the traditional boundaries between studies on motor skills and linguistics have been bridged by research on the role of the body in human communication (see Meteyard et al. 2012 for a review). A second contributing factor could be, as we have discussed here, that research on both grasping and gestures has often in the past been exceedingly near-sighted with its focus on tracing the linear development of a given skill, rather than on ways of analysing how context or co-occurring skills could play a relevant role. Finally, initial studies on SLs tended to focus on their discreet, arbitrary and categorical nature, which make them more like spoken languages, thereby overlooking iconic expressions in many SL structures (Perniss and Vigliocco 2014). More recent studies have begun considering gestures in terms of an essentially motoric activity and signs as enriched by highly iconic structures. Currently, not only have various studies shown that in SLs there is a large gestural component but also conversely, that more recent studies on gestures have begun to adopt many strategies for analysis borrowed from SLs studies (Vermeerbergen and Demey 2007; Boyes Braem and Bram 2000). While SL research originally tended to stress a distinction between signs and gestures, today researchers are not afraid to document an opposite tendency, recognizing the common sources of both signs and gestures. We hope that in the future a stronger integration between studies on motor skills, gesture research and sign linguistics will include a major focus on the role of the body and the environment in communication, thus hopefully allowing us all to fully capture the richness of human communication hidden in the shapes of our hands.

**Acknowledgements** This chapter is a part of a project that has received funding from the European Union's Horizon 2020 research and innovation programme under the Marie Skłodowska-Curie grant agreement No 660468 to LS. Portions of this work were inspired by thought provoking topics discussed during the ABLE Workshop "From Tools and Gestures to the Language-Ready Brain", Atlanta, GA, April 2010. Authors wish to thank two anonymous reviewers for useful comments on a previous version of this chapter.

# References

Acredolo, L., & Goodwyn, S. (1988). Symbolic gesturing in normal infants. *Child Development, 59,* 450–466.
Alibali, M. W., & Goldin-Meadow, S. (1993). Gesture-speech mismatch and mechanisms of learning: What the hands reveal about a child's state of mind. *Cognitive Psychology, 25,* 468–523.
Ann, J. (1996). On the relation between ease of articulation and frequency of occurrence of handshapes in two sign languages. *Lingua, 98,* 19–41.
Arbib, M. A., Gasser, B., & Barrès, V. (2014). Language is handy but is it embodied? *Neuropsychologia, 55,* 57–70.
Austin, J. L. (1962). *How to do things with words.* Cambridge, MA: Harvard University Press.
Bates, E., Benigni, L., Bretherton, I., Camaioni, L., & Volterra, V. (1979). *The emergence of symbols: Cognition and communication in infancy.* New York: Academic Press.
Bates, E., Camaioni, L., & Volterra, V. (1975). The acquisition of performatives prior to speech. *Merrill-Palmer Quarterley of Behavior and Development, 21*(3), 205–226.
Behne, T., Liszkowski, U., Carpenter, M., & Tomasello, M. (2012). Twelve-month- olds' comprehension and production of pointing. *British Journal of Developmental Psychology, 30,* 359–375.
Bernardis, P., Bello, A., Pettenati, P., Stefanini, S., & Gentilucci, M. (2008). Manual actions affect vocalizations in infants. *Experimental Brain Research, 184,* 599–603.
Bhat, A. N., Galloway, J. C., & Landa, R. J. (2012). Relation between early motor delay and later communication delay in infants at risk for autism. *Infant Behavior & Development, 35,* 838–846.
Boria, S., Fabbri-Destro, M., Cattaneo, L., Sparaci, L., Sinigaglia, C., Santelli, E., et al. (2009). Intention understanding and autism. *PLoSOne, 4*(5), e5596.
Boyatzis, C. J., & Watson, M. W. (1993). Preschool children's symbolic representation of objects through gestures. *Child Development, 64,* 729–735.
Boyes Braem, P. (1981). *Features of the handshape in American Sign Language.* Unpublished doctoral dissertation, University of California, San Diego.
Boyes Braem, P. (1994). Acquisition of handshape in American Sign Language: A preliminary analysis. In V. Volterra & C. J. Erting (Eds.), *From gesture to language in hearing and deaf children* (pp. 107–127). Washington, DC: Gallaudet University Press.

Boyes Braem & Bram. (2000). A pilot study of the expressive gestures used by classical orchestra conductors emmorey. In N. Mahwah (Ed.), *The Signs of Language Revisited, Karen & Lane, Harlan*. Jersey: Lawrence Erlbaum Associates.

Boyes Braem, P., Pizzuto, E., Volterra, V. (2002). The interpretation of signs by (Hearing and Deaf) members of different cultures. In R. Schulmeister, H. Reinitzer (Eds.), *Progress in sign language research. In Honor of Siegmund Prillwitz* (pp. 187–219). Hamburg: Signum-Verlag.

Brentari, D., Di Renzo, A., Keane, J., & Volterra, V. (2014). Cognitive, cultural and linguistic sources of a handshape distinction expressing agentivity. *Topics in Cognitive Science, 7*(1), 95–123.

Bruner, J. S. (1974). The ontogenesis of speech acts. *Journal of Child Language, 2*, 1–19.

Bruner, J. S. (1975). From communication to language—A psychological perspective. *Cognition, 3*(3), 255–287.

Butcher, C. & Goldin-Meadow, S. (2000). Gesture and the transition from one- to two-word speech: When hand and mouth come together. In D. McNeill (Ed.), *Language and gesture*, pp. 235–257). Cambridge: Cambridge University Press.

Butterworth, G. (2003). Pointing is the royal road to language for babies. In S. Kita (Ed.), *Pointing: Where language, culture, and cognition meet* (pp. 9–33). Mahwah, NJ: Lawrence Erlbaum Associates Publishers.

Butterworth, G., & Hopkins, B. (1988). Hand-mouth coordination in the new-born baby. *British Journal of Developmental Psychology, 6*, 303–314.

Capirci, O., Contaldo, A., Caselli, M. C., & Volterra, V. (2005). From actions to language through gestures. *Gesture, 5*(1/2), 155–177.

Capirci, O., Cristilli, C., De Angelis, V., & Graziano, M. (2011). Learning to use gesture in narratives: Developmental trends in formal and semantic gesture competence. In G. Stam & M. Ishino (Eds.), *Integrating gestures* (pp. 189–200). Amsterdam, Netherlands: Benjamins.

Capirci, O., Iverson, J., Pizzuto, E., & Volterra, V. (1996). Gesture and words during the transition to two-word speech. *Journal of Child Language, 23*, 645–673.

Carpenter, M., Nagell, K., & Tomasello, M. (1998). Social cognition, joint attention, and communicative competence from 9 to 15 months of age. *Monographs of the Society for Research in Child Development, 63*(4), 1–143.

Carter, M. (1981). The acquisition of British Sign Language (BSL: A first analysis), Unpublished manuscript.

Caselli, M. C. (1990). Communicative gestures and first words. In V. Volterra, & C. J. Erting (Eds.), *From gesture to language in hearing and deaf children* (pp. 56–67). Berlin/New York: Springer-Verlag.

Caselli, M. C., Rinaldi, P., Stefanini, S., & Volterra, V. (2012). Early action and gesture "Vocabulary" and its relation with word comprehension and production. *Child Development, 83*(2), 526–542.

Church, R. B., & Goldin-Meadow, S. (1986). The mismatch between gesture and speech as an index of transitional knowledge. *Cognition, 23*(1), 43–71.

Colletta, J. M. (2004). *Le développment de la parole chez l'enfant agé de 6 à 11 ans: corps, language et cognition*. Sprimont, Belgique: Mardaga.

Conlin, K. E., Mirus, G. R., Mauk, C., & Meier, R. P. (2000). The acquisition of first signs: place, handshape and movement. In C. Chamberlain, J. P. Morford, & R. I. Mayberry (Eds.), *Language acquisition by eye* (pp. 51–69). Mahwah, NJ: Erlbaum.

Connolly, K., & Dalgleish, M. (1989). The emergence of a tool-using skill in infancy. *Developmental Psychology, 25*(6), 894–912.

Connolly, K., & Elliott, J. (1972). The evolution and ontogeny of hand function. In N. Blurton-Jones (Ed.), *Ethological studies of child behaviour* (pp. 329–381). Cambridge: Cambridge University Press.

Cushing, F. H. (1892). Manual concepts: A study of the influence of hand-usage on culture-growth. *The American Anthropologist, 5*(4), 289–318.

Cutkosky, M. R. (1989). On grasp choice, grasp models, and the design of hands for manufacturing tasks. *IEEE Transactions on Robotics and Automation, 5*(3), 269–279.

De Vries, J. I. P., Visser, G. H. A., & Prechtl, H. F. R. (1982). The emergence of fetal behavior. I Qualitative aspects. *Early Human Development, 7*, 301–322.

Deacon, T. W. (1997). *The symbolic species: The co-evolution of language and the brain*. W.W. Norton: New York.

Dick, A. S., Overton, W. F., & Kovacs, S. L. (2005). The development of symbolic coordination: Representation of hand-as-hands, executive function, and theory of mind. *Journal of Cognition and Development, 6*, 133–161.

Eccarius, P., & Brentari, D. (2008). Handshape coding made easier: A theoretically based notation for phonological transcription. *Sign Language & Linguistics, 11*(1), 69–101.

Fenson, L., Marchman, V. A., Thal, D. J., Dale, P. S., Reznick, J. S., & Bates, E. (2007). *The MacArthur communicative development inventories: User's guide and technical manual* (2nd ed.). Baltimore, MD: Brookes.

Fogel, A., & Hannan, T. E. (1985). Manual actions of nine- to fifteen-week-old human infants during face-to-face interactions with their mothers. *Child Development, 56*, 1271–1279.

Franco, F. & Butterworth, G. (1996). Pointing and social awareness: Declaring and requesting in the second year. *Journal of Child Language, 23*(20), 307–336.

Gallagher, S. (2005). *How the body shapes the mind*. Oxford: Oxford University Press.

Gentilucci, M., Dalla Volta, R., & Giannelli, C. (2008). When the hands speak. *Journal of Physiology—Paris, 102*, 21–31.

Geraci, C. (2012). Language policy and planning: The case of Italian sign language. *Sign Language Studies, 12*(4), 494–518.

Gesell, A., & Ilg, F. L. (1937). *Feeding behavior of infants*. Philadelphia, PA: Lippincott.

Gibson, J. J. (1979). *The ecological approach to visual perception*. Boston: Houghton-Mifflin.

Goldin-Meadow, S. (2017). Using our hands to change our minds. *Wiley Interdisciplinary Reviews: Cognitive Science, 8*(1–2).

Goldin-Meadow, S., & Morford, M. (1990). Gesture in early child language. In Volterra, V. & Erting, C. J. (Eds.), *From gesture to language in hearing and deaf children* (pp. 249–262). Berlin/New York: Springer-Verlag.

Grover, L. (1988). *Comprehension of the pointing gesture in human infants*. Unpublished doctoral dissertation, Southampton, England: University of Southampton.

Gullberg, M., de Boot, K., & Volterra, V. (2008). Gestures and some key issues in the study of language development. *Gesture, 8*(2), 149–179.

Gurney, D. J., Pine, K. J., & Wiseman, R. (2013). The gestural misinformation effect: Skewing eyewitness testimony through gesture. *The American journal of psychology, 126*(3), 301–314.

Halverson, H. M. (1943). The development of prehension in infants. In R. G. Barker, J. S. Kounin, & Wright, H. F. (Eds.), *Child development and behaviour*, London: McGraw-Hill.

Iverson, J. M. (2010). Developing language in a developing body: the relationship between motor development and language development. *Journal of Child Language, 37*(2), 229–261.

Iverson, J., & Thelen, E. (1999). Hand, mouth and brain. The dynamic emergence of speech and gestures. *Journal of Consciousness Studies, 6*(11–12), 19–40.

Karnopp, L. B. (2002). Phonology acquisition in Brazilian Sign Language. In G. Morgan & B. Woll (Eds.), *Directions in sign language acquisition* (pp. 29–53). Amsterdam: John Benjamins Publishing Company.

Kendon, A. (2004). *Gesture: visible action as utterance*. Cambridge: Cambridge University Press.

Kisch, D. (1996). Adapting the environment instead of oneself. *Adaptive Behavior, 4*(3/4), 415–452.

Leung, E. H., & Rheingold, H. L. (1981). Development of pointing as a social gesture. *Developmental Psychology, 17*, 215–220.

Libertus, K., Sheperd, K. A., Ross, S. W., & Landa, R. L. (2014). Limited fine motor and grasping skills in 6-month-old infants at high risk for autism. *Child Development, 85*(6), 2218–2231.

Lifter, K., & Bloom, L. (1989). Object knowledge and the emergence of language. *Infant Behavior and Development, 12*, 395–423.

Malafouris, L. (2013). *How things shape the mind. A theory of material engagement*. Cambridge, MA: MIT Press.

Marentette, P., Pettenati, P., Bello, A., & Volterra, V. (2016). Gesture and symbolic representation in Italian and English-Speaking Canadian 2-year-olds. *Child Development, 87*(3), 944–961.

Masataka, N. (2003). From index-finger extension to index-finger pointing: Ontogenesis of pointing in preverbal infants. In S. Kita (Ed.), *Pointing: where language, culture, and cognition meet* (pp. 69–109). Mahwah, NJ, US: Lawrence Erlbaum Associates Publishers.

McCarty, M. E., Clifton, R. K., & Collard, R. R. (2001). The beginnings of tool use by infants and toddlers. *Infancy, 2*(2), 233–256.

McIntire, M. (1977). The acquisition of American Sign Language hand configurations. *Sign Language Studies, 16,* 247–266.

McNeill, D. (1992). *Hand and mind. What the hands reveal about thought.* Chicago: University of Chicago Press.

Meteyard, L., Cuadrado, S. R., Bahrami, B., & Vigliocco, G. (2012). Coming of age: A review of embodiment and the neuroscience of semantics. *Cortex, 48,* 788–804.

Morgan, G., Barrett-Jones, S., & Stoneham, H. (2007). The first signs of language: Phonological development in British Sign Language. *Applied Psycholinguistics, 28,* 3–22.

Morissette, P., Ricard, M., & Decarie, T. G. (1995). Joint visual attention and pointing in infancy—A longitudinal study of comprehension. *British Journal of Developmental Psychology, 13,* 163–175.

Napier, J. R. (1956). The prehensile movements of the human hand. *Journal of Bone and Joint, 38-B*(4), 902–13.

Napier, J. R. (1980). *Hands.* Princeton, NJ: Princeton Science Library.

Newell, K. M. (1986). Constraints on the development of coordination. In M. G. Wade & H. T. A. Whiting (Eds.), *Motor development in children: Aspects of coordination and control* (pp. 341–360). Boston: Martinus Nijhoff.

Newell, K. M., Scully, D. M., McDonald, P. V., & Baillargeon, R. (1990). Task constraints and infant grip configurations. *Developmental Psychobiology, 22*(8), 817–832.

O'Reilly, A. W. (1995). Using representations: Comprehension and production of actions with hand-as-hands. *Child Development, 66,* 999–1010.

Overton, W. F., & Jackson, J. P. (1973). The representation of hand-as-hands in action sequences: A developmental study. *Child Development, 44,* 309–314.

Ozonoff, S., Young, G. S., Carter, A., Messinger, D., Yirmiya, N., Zwaigenbaum, L., et al. (2011). Recurrence risk for autism spectrum disorders: A baby siblings research consortium study. *Pediatrics, 128*(3), e488–e495.

Perniss, P., & Vigliocco, G. (2014). The bridge of iconicity: From a world of experience to the experience of language. *Philosophical Transactions of the Royal Society, B, 369*(1651), 20130300.

Pettenati, P., Sekine, K., Congestri', E., & Volterra, V. (2012). A comparative study on representational gestures in Italian and Japanese children. *Journal of Nonverbal Behavior, 36*(2), 149–164.

Pettenati, P., Stefanini, S., & Volterra, V. (2009). Motoric characteristics of representational gestures produced by young children in a naming task. *Journal of Child Language, 37*(4), 887–911.

Pietrandrea, P. (1997). I dizionari della LIS: analisi qualitativa e quantitativa. In M. C. Caselli & S. Corazza (Eds.), *LIS. Studi, esperienze e ricerche sulla lingua dei segni in Italia* (pp. 255–259). Pisa: Edizioni del Cerro.

Pietrandrea, P. & Russo, T. (2007). Diagrammatic and Imagic Hypoicons in signed and verbal languages. In E. Pizzuto, P. Pietrandrea, R. Simone (Eds.), *Verbal and signed languages. Comparing structures, constructs and methodologies* (pp. 35–56). Berlin: Mouton de Gruyter.

Pine, K. J., Lufkin, N., & Messer, D. (2004). More gestures than answers: Children learning about balance. *Developmental Psychology, 40*(6), 1059–1067.

Pizzuto, E., Cameracanna, E., Corazza, S., & Volterra, V. (1995). Terms for spatio-temporal relations in Italian Sign Language (LIS): What they can tell us about iconicity in sign and speech. In R. Simone (Ed.), *Iconicity in language* (pp. 237–256). New York-Amsterdam: Benjamins.

Pizzuto, E., & Corazza, S. (1996). Noun morphology in Italian Sign language (LIS). *Lingua, 98,* 169–196.

Radutzky, E. (1992). *Dizionario bilingue elementare della Lingua Italiana dei Segni*. Rome: Ediziona Kappa.

Rogers, S. J. (2009). What are infant siblings teaching us about autism in infancy? *Autism Research, 2,* 125–137.

Russo Cardona, T., & Volterra, V. (2007). *Le lingue dei segni. Storia e Semiotica*. Roma: Carocci.

Searle, J. (1969). *Speech acts: An essay in the philosophy of language*. Cambridge, England: Cambridge University Press.

Sparaci, L. (2010). Discourse and action: analyzing the possibility of a structural similarity. In M. D'Agostino, G. Giorello, F. Laudisa, T. Pievani, & C. Sinigaglia (Eds.), *SILFS—New essays in logic and philosophy of science* (pp. 493–504). London: College.

Sparaci, L., Northurp, J., Capirci, O. & Iverson, J. M. (forthcoming) From grasping tools to grasping words in infants at high-risk for Autism Spectrum Disorders. *Journal of Autism and Developmental Disorders.*

Sparaci, L., Stefanini, S., D'Elia, L., Vicari, S., & Rizzolatti, G. (2014). What and why understanding in autism spectrum disorders and Williams syndrome: Similarities and differences. *Autism Research, 7*(4), 421–432.

Sparaci, L., Stefanini, S., Marotta, L., Vicari, S., & Rizzolatti, G. (2012). Understanding motor acts and motor intentions in Williams syndrome. *Neuropsychologia, 50*(7), 1639–1649.

Stefanini, S., Bello, A., Volterra, V., & Carlier, M. (2008). Types of prehension in children with Williams-Beuren syndrome: A pilot study. *European Journal of Developmental Psychology, 5* (3), 358–368.

Stokoe, W. C. (1960). Sign language structure: An outline of the visual communication systems of the American deaf. *Studies in Linguistics, Occasional Papers 8. Surgery, 38B,* 902–913.

Sutton-Spence, R., & Woll, B. (1999). *The linguistics of British Sign language: An introduction.* Cambridge: Cambridge University Press.

Twitchell, T. (1965). The automatic grasping responses of infants. *Neuropsychologia, 3,* 247–259.

Twitchell, T. (1970). Reflex mechanisms and the development of prehension. In K. Connolly (Ed.), *Mechanisms of motor skill development* (pp. 25–45). New York: Academic Press.

Vermeerbergen, M., & Demey, E. (2007). Sign + Gesture = Speech + Gesture? Comparing aspects of simultaneity in Flemish Sign Language to instances of concurrent speech and gesture. In M. Vermeerbergen, L. Leeson, & O. Crasborn (Eds.), *Simultaneity in signed languages: Form and function* (pp. 257–282). Amsterdam: Johns Bejamins Publishing Company.

Visalberghi, E., & Tomasello, M. (1989). Primate causal understanding in the physical and psychological domains. *Behavioural Processes, 42,* 189–203.

Volterra, V. (Ed.). (1987). *La Lingua Italiana dei Segni La comunicazione visivo-gestuale dei sordi*. Bologna: Il Mulino. (Nuova Edizione 2004).

Volterra, V., Capirci, O., Caselli, M. C., Rinaldi, P., & Sparaci, L. (2017). Developmental evidence for continuity from action to gesture to sign/word. *Language Interaction and Acquisition, 8,*13–42.

Von Tetzchner, S. (1984). First signs acquired by a Norwegian deaf child with hearing parents. *Sign Language Studies, 44,* 225–257.

Wilson, F. (1998). *The hand*. New York: Vintage Books.

Zinober, B., & Martlew, M. (1985). Developmental changes in four types of gesture in relation to acts of vocalizations from 10 to 21 months. *British Journal of Developmental Psychology, 3,* 293–306.

# Primates' Propensity to Explore Objects: How Manual Actions Affect Learning in Children and Capuchin Monkeys

**Fabrizio Taffoni, Eugenia Polizzi di Sorrentino, Gloria Sabbatini, Domenico Formica and Valentina Truppa**

**Abstract** Humans and other animals have a strong propensity to explore the environment. When human infants, as well as other primates, face the opportunity to interact with the environment by manipulating objects, they may discover and learn the contingency between one action and its outcome. Thus, manipulation, as a form of spontaneous exploration, has a great biological significance, since it allows to discover and learn the relationship between action and effect, enabling humans and other animals to plan goal-directed tasks. How do the specific characteristics of the primate's body influence this process? With its large amount of degrees of freedom, sensors, and nervous terminations, the hand is the main interface with the external world, and it profoundly influences the primates' interaction with the environment. How does object exploration mediated by manual actions affect the acquisition of problem-solving abilities? To try to answer this question, we experimentally compared how children and capuchin monkeys (*Sapajus* spp.)—nonhuman primates well known for their manual dexterity and for being curious and highly manipulative—acquire new cause–effect relations through spontaneous manual exploration of a new environment. The experiments were carried out with the mechatronic board, an innovative device specifically designed to allow inter-species comparative research. The board allowed testing whether spontaneous

---

F. Taffoni (✉) · D. Formica
Unit of Biomedical Robotics and Biomicrosystems, Department of Engineering,
Università Campus Bio-Medico di Roma, via Álvaro Del Portillo 21, 00128 Rome, Italy
e-mail: f.taffoni@unicampus.it

D. Formica
e-mail: d.formica@unicampus.it

E. Polizzi di Sorrentino · G. Sabbatini · V. Truppa
Unit of Cognitive Primatology and Primate Center, ISTC-CNR, Rome, Italy
e-mail: eugenia.polizzi@istc.cnr.it

G. Sabbatini
e-mail: gloria.sabbatini@istc.cnr.it

V. Truppa
e-mail: valentina.truppa@istc.cnr.it

manipulation of objects (not instrumental to achieve any specific goal) improved subjects' ability to solve a subsequent goal-directed task by retrieving the knowledge learned during previous exploration.

**Keywords** Object manipulation · Curiosity driven learning · Action-outcome contingency · Mechatronic board · Behavioral analysis

# 1 Object Manipulation from Manual Dexterity to Proclivity to Explore

The capability to interact with the environment and exploring the world around us is the key ingredient for the development of the brain and consequently for what is commonly considered *intelligence*. In fact, in the era of intelligent machines where huge investments of the leading high-tech companies are devoted to bringing what is called *artificial intelligence* (AI) to a new level of independence and versatility, most experts reckon that, to quote Jean-Christophe Baillie:

> there is no AI without robotics: that is because you need a body to interact with the world, and you need to interact with the world to develop your own, internal meaning for everything you encounter, hear, say, and do (Baillie 2016).

This vision is strongly related to so-called *sensorimotor theory*, mainly proposed by O'Regan and Noë (2001): It argues that the knowledge and the consciousness of a sensory experience, instead of being generated somewhere in the brain, are constituted by the set of objective laws concerning the interaction with the world that the experience involves, i.e., the objective laws linking actions to resulting sensory changes, called *sensorimotor contingencies* or *sensorimotor dependencies*. This process starts with becoming aware of one's own body and of how one can control it to produce desired movements and thus to affect the observed world: This seems to be essential to build one's own notion of space, distance, color, and all the other properties that the brain consciously and unconsciously perceives.

In primates, and humans in particular, the main actor of this exploration is the hand. It is a versatile organ that is used for grasping heavy or delicate objects and for performing highly complex manipulations based on fine motor control and precise sensory feedback. This is achieved by combining and orchestrating a large number of degrees of freedom, proprioceptive and exteroceptive sensors, a variety of levels of strength/velocity production, and a complex hierarchical motor control (Kapandji 1987). Each hand has as many controllable degrees of freedom (twenty-one) as both arms, wrists, and one leg combined. As the majority of manual tasks involve effort using more than one finger, the resulting possible combination of manipulative actions is extremely large (Cutkosky 1989).

In general, manual dexterity represents a hallmark of primate species and is one of the most commonly examined aspects in comparative studies (Fragaszy and Crast 2016; Preuschoft and Chivers 2012). How much different species converge or diverge with regard to the processes by which they come to understand the world allows us to investigate the evolution of mind and brain in living species. As such, comparative research on manipulative behaviors is highly informative on the evolution of both motor and cognitive skills of human and nonhuman primates. Primates acquire a rich repertoire of actions for manipulating objects during the course of their ontogeny, and learning how to exploit the information acquired through visual–manual exploration is a crucial developmental challenge.

According to Napier's seminal work (1956), the hand movements can be divided into two main categories: (1) prehensile movements or movements in which an object is seized and held partly or wholly within the compass of the hand, and (2) non-prehensile movements or movements in which no grasping or seizing is involved but by which objects can be manipulated by pushing or lifting motions of the hand as a whole or of the digits individually.

In this chapter, we mainly focus on the second class of hand movements to investigate if the sensorimotor contingencies based on manual exploratory behavior play a role in promoting the acquisition of new skills. In order to study how exploration affects learning new skills from an ontogenetic as well as from a phylogenetic approach, we compared explorative behaviors of 3- to 4-year-old children with those of capuchin monkeys, nonhuman primates well known for their manual dexterity, and their propensity to explore objects (Fragaszy et al. 2004). To objectively assess their interactions, we used an experimental setup, hereafter called mechatronic board, which has been specifically instrumented with sensors for recording and analyzing the aforementioned sensorimotor contingencies in this comparative perspective.

## 1.1 The Emergence of Exploratory Actions in Children

Over the course of the first year of life, human infants rapidly acquire skills in object-directed reaching and postural control (Cioni and Giuseppina 2013). These skills underlie increasingly sophisticated object exploration behaviors that yield information not only about objects in the world, but also about the effects of the infant's actions on those objects. Object exploration and the knowledge that it generates (in terms of reaching behavior and postural control) are, in turn, at the bases of later development of complex skills ranging from advanced motor planning skills to communication and language (Iverson 2010). Conversely, delays and impairments in reaching and postural development may negatively impact object exploration and consequently the aforementioned skills (Focaroli et al. 2015).

The emergence of reaching, the achievement of independent sitting, the elaboration of finely tuned object exploration, the development of increasingly sophisticated motor planning skills, and the growth of language and communication are among the most important milestones in the first few years of life.

All these skills are necessary for exploring the environment, and their development is strongly interrelated; for example, postural development and increased stability control of head and trunk are a prerequisite for skilled reaching movements and manipulation capabilities. It starts with the ability to maintain head position at midline (Spencer et al. 2000), which stabilizes the visual field, allowing infants to focus on targets toward which reaching will eventually be directed. Increased postural stability is related to better reaching performance (Fallang et al. 2000), while progression from supine to sitting creates a more biomechanically supportive context for arm movement (Out et al. 1998). In fact, in the seated posture, the gains in limb stability directly affect infant reaching and object exploration. In supine posture, arm movements are more effortful and less easily controlled and infants must constantly work against gravity to hold an object within the line of sight. When seated, however, infants are free to move hands and arms in less biomechanically challenging ways and the upright head position enlarges the field of view and stabilizes gaze, thereby promoting eye–hand coordination (Bertenthal and Von Hofsten 1998). Moreover, when infants acquire the ability to sit independently (self-sit), the hands are no longer needed for support and are thus free to move. Possibilities for object exploration are therefore further enhanced (Rochat and Goubet 1995).

## 1.2 The Capuchin Monkeys: Highly Dextrous Species

Among nonhuman primate species, tufted capuchins (genus *Sapajus*) are especially suitable for the comparative study of learning mediated by manual actions (Lynch Alfaro et al. 2012a, b). They are highly dextrous New World monkeys (see Fig. 1) with a strong propensity to explore the environment and manipulate objects (Fragaszy et al. 2004). Capuchins possess a relatively dense substrate of direct corticospinal motoneurons innervating individual digits of the hand (Bortoff and Strick 1993). In addition, their forearm muscles devoted to thumb motion show

**Fig. 1** A young wild bearded capuchin monkey (*Sapajus libidinosus*) is processing a tuberous root after having extracted it from the ground (*Photo* V. Truppa)

important parallels with those of chimpanzees and humans (Aversi-Ferreira et al. 2011). Thanks to these neuroanatomical adaptations, capuchins are able to perform independent movements of the digits and exert strong thumb manipulative forces, both characteristics which allow them to grasp objects with a variety of precision (Christel and Fragaszy 2000; Costello and Fragaszy 1988; Spinozzi et al. 2007; Spinozzi et al. 2004) and power grips (Truppa et al. 2016). Moreover, in contrast to other New World monkeys, capuchins possess well-differentiated posterior parietal cortical areas associated with (i) proprioception (area 2) and (ii) motor planning and internal body coordinates for visually guided reaching, grasping, and manipulation (area 5) (Padberg et al. 2007). Both wild and captive capuchins are known for their natural proclivity to manipulate objects and exploit resources that need to be extracted from a substrate (Fragaszy et al. 2004; Perry and Manson 2008; Spinozzi et al. 2007; Spinozzi et al. 1998; Terborgh 1983; Visalberghi et al. 2015a). They spontaneously use their hands to perform (i) object–object and object–surface combinations (Byrne and Suomi 1996; Fragaszy and Adams-Curtis 1991; Fragaszy and Boinski 1995; Panger 1998; Visalberghi 1988), (ii) object actions that are directed toward the self (e.g., eating food items from a dowel) (Sabbatini et al. 2016; Truppa et al. 2016; Zander and Judge 2015; Zander et al. 2013), and (iii) object actions that are directed toward external targets (e.g., using tools to obtain out-of-reach food items) (Visalberghi and Fragaszy 2006; Visalberghi et al. 2015b). Overall, capuchins' skills to manipulate and combine objects may improve their ability to discover and learn the relation between their actions and subsequent outcomes, thus making this species a good candidate to study action–outcome contingencies through manual actions in a comparative perspective.

## 2 Experiments on Children and Monkeys

In this section, we will present the experimental apparatus used and we will describe the protocol of our study.

### 2.1 The Mechatronic Board

A behavioral study that tackles how manipulation and outcome contingency affect learning in children and animal models requires a research frame able to (i) measure behaviors; (ii) provide a controlled sensory feedback in response to the action performed; and (iii) define a common metric to compare data from different species. In this respect, mechatronic platforms may represent a good solution since they allow to structure the environment using a common frame where children and nonhuman animals can be observed and assessed. Since these platforms can be provided with different instrumented objects for automatic manipulation monitoring, they make a real-time assessment of actions possible, enabling a reliable and repetitive delivery of well-defined stimuli according to the actions performed.

The platform designed for this study is composed of (i) a planar base, which can be equipped with instrumented interchangeable objects (mechatronic modules) stimulating different kinds of manipulative behaviors and allowing to record several kinds of actions (e.g., rotations, pushing, pulling, repetitive hand movements, button pressing); (ii) a frontal unit with three boxes, each with a sliding door controlled in a reprogrammable way by the actions performed on the modules (reward-releasing mechanism); and (iii) a system for visual and audio feedback delivery (from both the planar and the frontal base). Its functions can be described by a three-layer hierarchical architecture (see Fig. 2):

1. The physical level, made of the physical interfaces subjects, can directly interact with: push buttons, instrumented objects (mechatronic modules), and reward-releasing mechanisms.
2. The microcontroller-based middleware level manages the low-level I/O communication with the mechatronic modules and push buttons, the reward-releasing mechanism, and audio–visual stimuli.
3. The high-level control and supervision system, running on a remote laptop, manages and supervises the acquisition and the arbitrary association between actions and outcomes. The communication between the remote laptop and the

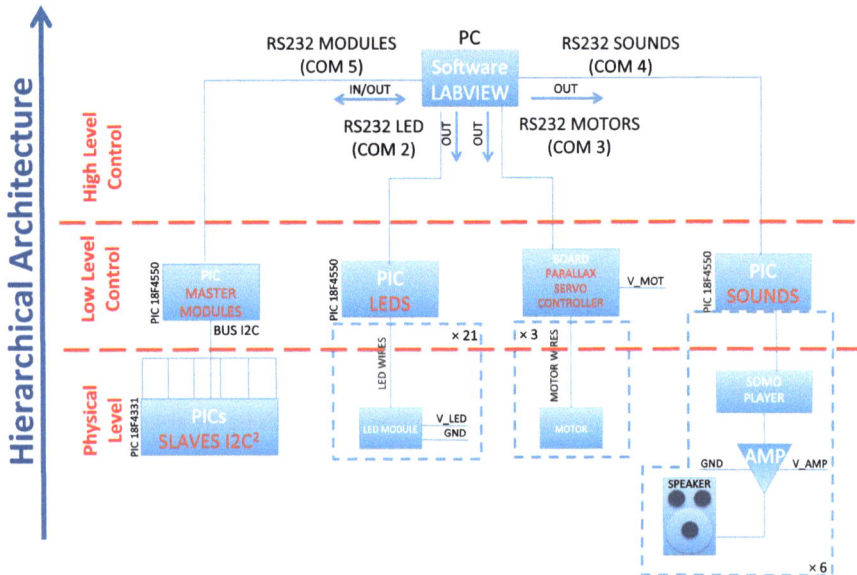

**Fig. 2** Hierarchical architecture of the board: physical level made by the interfaces with subject; local low-level control microcontroller based; high-level control running on a remote laptop

microcontroller-based level is provided by four different serial RS232 connections.

Two versions of the board have been designed: one for children and one for monkeys (Taffoni et al. 2013). They are slightly different so as to allow for interspecies differences. The monkey version of the board is heavier, bigger, and made of waterproof (since monkeys could urinate on it) and non-varnished materials (since they like to remove the paint with their teeth or nails), which are robust enough to contrast some typical monkeys' actions such as hitting, scratching, and biting. The children version is similar, but scaled in size and mainly made of wood (see Fig. 3).

The optimal board configuration for each species has been tailored after some pilot experiments on both children (Taffoni et al. 2012a) and monkeys (Taffoni et al. 2012b). In particular, the pilot with children suggested that we focus our investigation on 3- and 4-year-old children using simple push button modules. Indeed, children younger than 3 years of age were not able to keep their attention focused on the board for the necessary length of time without the intervention of the experimenter, while children older than 5 found the task boring, so much so that they did not perform it. Moreover, since children develop their manipulation skills in this age range (Focaroli et al. 2015), using simpler module allowed to avoid biases due to difficulties in manipulation related to motor development. The pilot with capuchin monkeys, a species well known to be manipulative and destructive when dealing with objects and food items (Fragaszy et al. 2004), has been carried out mainly to test the functioning of the board and its appropriateness for monkeys.

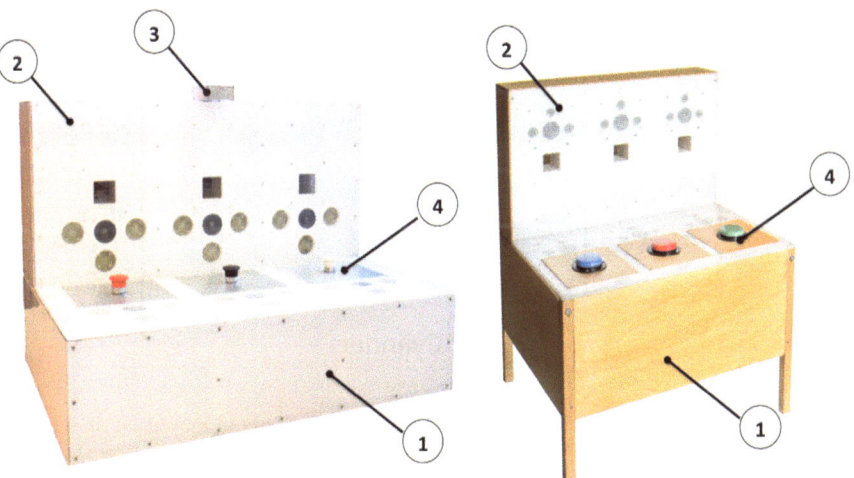

**Fig. 3** Mechatronic board for monkeys (*left*) and children (*right*): (1) planar base, (2) reward-releasing unit; (3) local wide-angle camera (only in the monkey board); (4) mechatronic modules. The stimuli/reward system is not visible to the subjects, and it controls the aperture and closure of the reward boxes as well as the visual and acoustic stimuli

Also, in this case, the board was equipped with simple push buttons. Data suggested that the buttons did not particularly interest capuchin monkeys per se. For this reason, we eventually tested the board equipped with three mechatronic modules: the circular tap, the fixed prism, and the 3-DOF cylinder (see Fig. 4). Modules have different numbers of affordances and mimicking actions usually performed in nature. The pilot showed that there was a tendency to perform fewer errors when

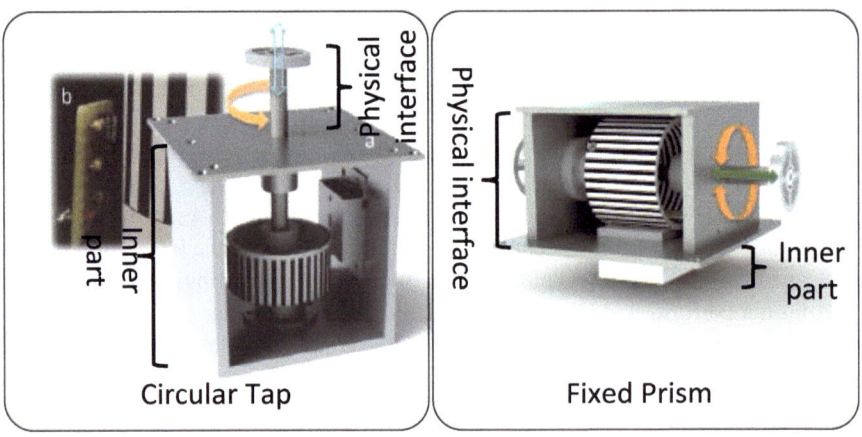

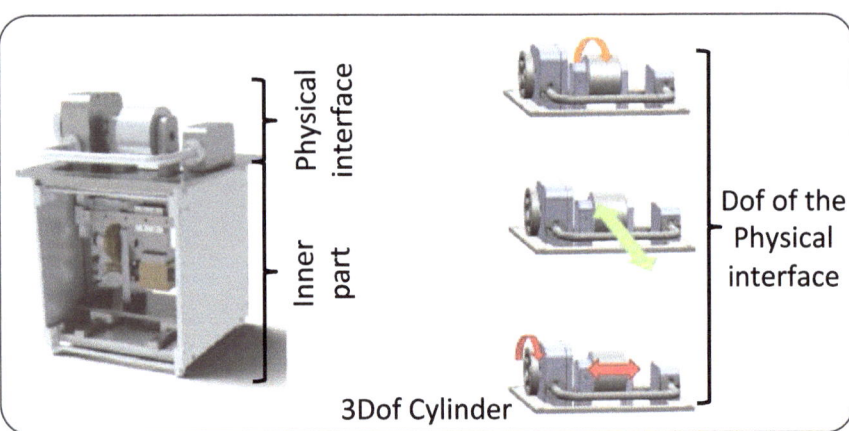

**Fig. 4** Mechatronic modules. The arrows represent the possibilities of movement. In the *upper left panel* Circular Tap, overall layout and details of the encoder electronics (on the *right*). This module allows two actions, suggested by two different affordances: rotation and lift up/down. *Upper right panel* Fixed Prism, the frontal wall has been removed allowing to see inner mechanism. This module allows two actions: horizontal rotation and translation. Bottom panel: 3-DOF cylinder. On the left, the overall layout is represented; on the right degrees of freedom: central wheel rotation (up); horizontal translation (*middle*); translation of the central wheel driven by lateral circular handle bottom

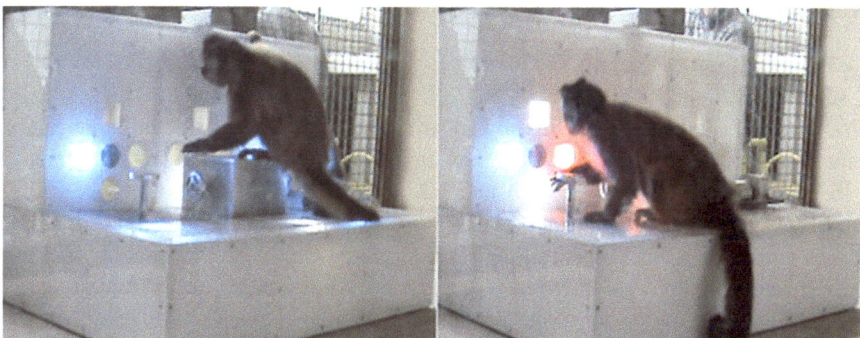

**Fig. 5** Example of a typical behavior observed during the pilots with monkeys. On the *left*, the animal spent a big portion of time during the interaction with the board up to the central module. On the *right*, a typical example of manipulative behavior

manipulating a two-affordance module compared to the three-affordance ones (3-DOF cylinder), suggesting that the complexity of the modules could negatively influence individual performance. Moreover, we observed that animals preferred to stay in the central part of the planar base (see Fig. 5 left) performing fewer actions on the lateral modules (see Fig. 5 right). The pilot experiments led us to set up the two mechatronic boards as follows: (i) for children, the platform was equipped with three simple round push buttons (diameter 60 mm), a blue button (BB) on the left, a red button (RB) in the center, and a green button (GB) on the right (all three reward-releasing mechanisms were used); (ii) for monkeys, the platform was equipped with two circular taps, one on the right side of the platform and one on the left side. The central part of the planar base was left without any module. To avoid monkeys' behavior to be biased by spatial proximity (e.g., associating modules with their respecting frontal boxes), only the central reward-releasing mechanism was used, while the lateral ones were closed by a white Plexiglas shield. These two different configurations allowed on the one hand to promote the explorative actions mostly used by children and monkeys, and on the other hand, they represented a trade off between novelty and complexity, necessary to promote spontaneous exploration without discouraging subjects (Kaplan and Pierre-Yves 2007). In this chapter, we reviewed the results of two studies carried out in parallel on children (Taffoni et al. 2014) and monkeys (Polizzi di Sorrentino et al. 2014) in order to have a comparative view on how manipulation and outcome contingency affect the way children and monkeys explore a new environment and how it enables the acquisition and retention of new skills.

## 2.2 Participants

**Children**

We enrolled two groups of different mean age to test if there was any age-related effect in the way children explore a new environment: 12 children were 3 years old ($36.7 \pm 0.8$ months, mean $\pm$ SD) and 12 children were 4 years old ($48.7 \pm 0.8$). Subjects were recruited from two day-care centers: the Università Campus Bio-Medico di Roma institutional kindergarten and the Casa dei Bambini Montessori Kindergarten in Rome. Children were individually tested in a quiet and familiar room of the center in the presence of their teachers. The children's parents signed a written informed consent which described the purpose of the experiment.

**Monkeys**

Sixteen socially housed adult tufted capuchin monkeys (eight females and eight males) hosted at the Unit of Cognitive Primatology and Primate Centre, ISTC-CNR of Rome, were enrolled in the study. The monkeys lived in four groups housed in enclosures consisting of an outdoor area and two indoor cages (experimental area). Capuchins were tested individually in the indoor area, which they accessed through a sliding door from the adjacent outdoor enclosure. Each subject was separated from the group for the sole purpose of testing, just before each testing session.

## 2.3 Procedure

The experiments consisted of three phases administered in the following order: (1) familiarization phase, (2) free manipulation phase, and (3) test phase. Both children and monkeys were randomly divided into two groups: experimental (EXP) and control (CTRL) group. Each subject of the CTRL group was paired with one subject of the EXP group, matched according to age, sex, and possibly manipulation skills. Here below we describe how the procedure was implemented in both species.

### 2.3.1 Familiarization Phase

The goal of this phase was to estimate the starting skills of subjects and their interest in exploring the board. The familiarization phase for both children and monkeys lasted 5 min.

**Children**

During this phase, children were presented with a version of the board in which only the audio–visual stimuli coming from the planar base were activated: Whenever a button was pressed, the lights above it switched on and a xylophone sound was produced (three different tones corresponded to the three different buttons).

### Monkeys

In order to limit the possible decrease in interest due to habituation to stimuli, capuchin monkeys were presented with a version of the board equipped only with one central metal handle (i.e., a different object compared to the circular taps used in the following phases) that could be rotated. No visual or audio stimuli were associated with the rotation of the handle.

### 2.3.2 Free Manipulation Phase

During this phase, subjects were allowed to interact with the full version of the board and to freely explore it. In both experiments, for the EXP group the board was programmed so as to respond contingently to the manipulation of subjects by producing specific sounds and lights (see below). By contrast, CTRL subjects could operate on the modules, but no outcome was directly produced. Instead, the outcomes they experienced were identical to those performed by their paired EXP subjects (yoked condition). This was done to provide CTRL subjects with the same number of outcomes as their paired EXP subjects, while preventing the former from repeatedly experiencing congruent associations between their own actions and outcomes.

### Children

For the EXP subjects, the board was programmed to respond to any pressing of the buttons with contingent visual and auditory stimuli and to open a specific box when its associated button was kept pressed for more than one second. A simple push (SP), i.e., a button pressed for less than 1 s, switched on the lights above the button (on the planar base) and produced a xylophone sound. An extended push (EP), i.e., when the button was pressed for one second or more, produced the same stimuli as an SP from the planar base, but it also produced the opening of a box (always empty in this phase) and the corresponding visual and audio stimuli from the frontal unit: the interior of the box lit up, the lights above the box switched on, and the speaker near the box produced an animal sound (a different one for each box: a rooster's, a frog's, or a cat's call). The association between buttons and boxes was programmed to be direct (the button opens the box in front of it) or crossed (the button opens the box on its left or right side), see Fig. 6. The side (left, right) of the crossed association between buttons and boxes was counterbalanced among subjects.

### Monkeys

EXP subjects could manipulate the modules (and experience the relative action–outcome associations) and open the central box by performing a specific action. This action consisted of rotating the tap of one of the two modules of at least 45 degrees (either clockwise or counterclockwise). When the correct rotation was performed (box opening rotation), the box opened along with a specific sound, and a light stimulus appeared below and inside the box. The other actions (rotating the tap of the other module and performing lifting actions on both modules) did not open the box

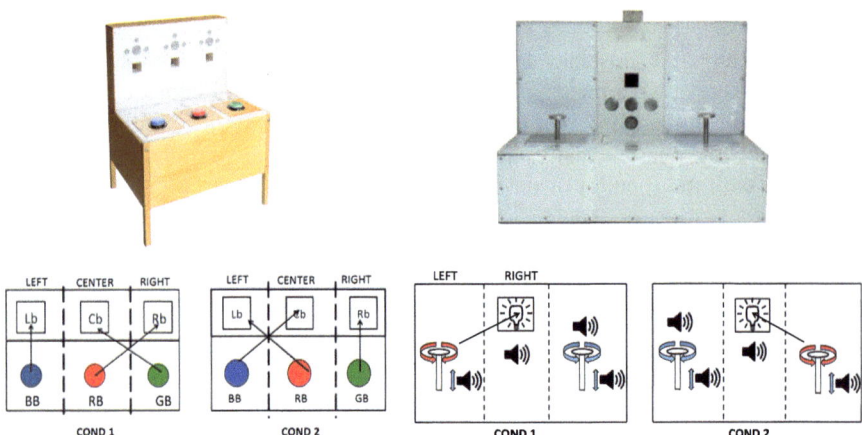

**Fig. 6** Mechatronic board: set up used with children, on the left; set up used with monkeys, on the right

and were only associated with sounds. The position of the module associated with the opening of the box (left, right) was counterbalanced among subjects.

### 2.3.3 Test Phase

While in the free manipulation phase the box(es) was/were always empty, in this phase we introduced a reward (respectively, a sticker for children and a peanut for monkeys) to promote goal-directed actions. During this phase, the outcomes produced (e.g., sounds, lights, box openings) depended on subject's actions, so both EXP and CTRL groups could experience the action–outcome contingencies. Relations between actions and outcomes were set to be the same as the ones proposed in the previous phase.

**Children**

The test phase consisted of nine consecutive trials, each lasting maximum 2 min. During each trial, the subject was asked to retrieve a sticker placed in one of the three boxes (the sticker was always visible as the box door was transparent). Three different sequences of the sticker position were used in order to avoid a bias effect due to the presentation order of the reward. The sequences were randomly assigned and counterbalanced among EXP subjects. Paired CTRL subjects received the same sequence order. To open the box and retrieve the sticker, children had to keep the correct button pressed for at least one second. When the subject succeeded in opening the door and getting the reward, a new reward was placed inside the next box of the sequence. If the subject did not get the reward within 2 min, the same reward was moved inside the following box.

## Monkeys

The test phase consisted of ten consecutive trials, each lasting maximum 2 min. For each of the ten trials, the experimenter baited the box with one unshelled peanut kernel. The peanut was always visible as the box door was transparent. In each trial, subjects could manipulate the modules, and if the correct action (i.e., the box-opening rotation) was performed, the box opened so that they could retrieve the reward, and a new trial started over. If the correct action was not performed within 2 min, the subject was attracted into the adjacent enclosure, and the next trial started over.

## 3 Results and Discussion

*The role of action–outcome contingencies in promoting manipulation*

In both studies, the starting skills of EXP and CTRL groups were assessed using the data collected during the familiarization phase. To select the most appropriate statistical test to be performed, the assumptions of normality and variance homogeneity of the relevant variables were verified each time. No significant differences were observed between the two groups, neither in children nor in monkeys, suggesting that individuals were well balanced among groups. During the free manipulation phase, four-year-old EXP children showed a significantly higher level of interaction (expressed as number of pushes) than CTRL ones (see Fig. 7). Within our monkey study, not only did the EXP subjects spend more time in contact with the board, but they also maintained a higher level of manipulation and board exploration throughout the entire free manipulation phase compared to CTRL subjects. By contrast, CTRL subjects in both monkeys and four-year-old children

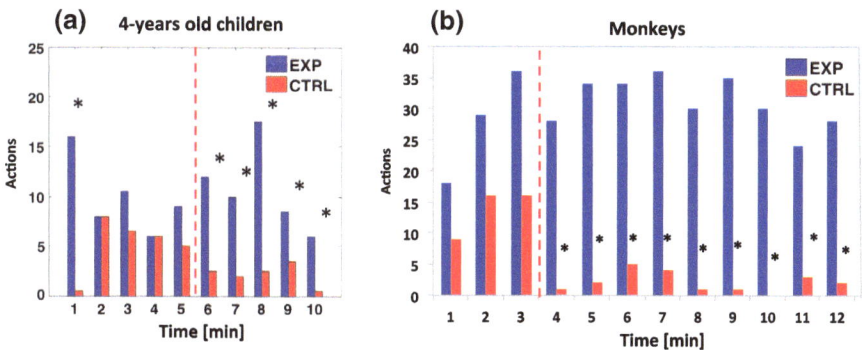

**Fig. 7** Mean number of actions performed during 1-min time bins of free manipulation phase in the 4-year-old children subset (**a**) and in the capuchin monkeys (**b**) study. Asterisks represent significant differences between EXP and CTRL subjects in each separate time bin

significantly dropped their interest toward manipulation at similar points in time (at about half of the free manipulation phase). These findings suggest that in both children and monkeys the opportunity to discover action–outcome contingencies was the strongest motivating aspect underlying manipulation, whose absence in the CTRL condition represented a constraint toward further exploring the board. Interestingly, no significant differences were found in the level of interaction between three-year-old EXP and CTRL children. This result suggests that, even in the absence of action contingency, the purely novel and surprising aspects of the stimuli may be strongly motivating for three-year-old subjects.

*The discovery of action–outcome contingencies through manipulation promotes knowledge of the environment*

The opportunity to learn the relation between action and outcome not only kept subjects' engagement high in the absence of rewards (e.g., free manipulation phase), but also increased their ability to solve a subsequent task, whose solution required recalling information gathered in the previous phase (e.g., recalling which action opened the box(es)—whether it was rotating the tap of the correct module in the monkey study, or pushing extensively the correct button in the children study). In both the children and the monkey study, the amount of effective actions performed during the free manipulation phase (e.g., those actions that are effective in opening the box in the absence of reward) correlated negatively with the time needed to retrieve the reward in the subsequent test phase (Fig. 8a, b). Specifically, the higher the percentage of extended pushes performed in the free manipulation phase, the shorter the time needed by children to retrieve the reward in the test phase. Similarly, EXP monkeys that more frequently performed the effective rotation during the free manipulation phase showed a shorter latency to solution in

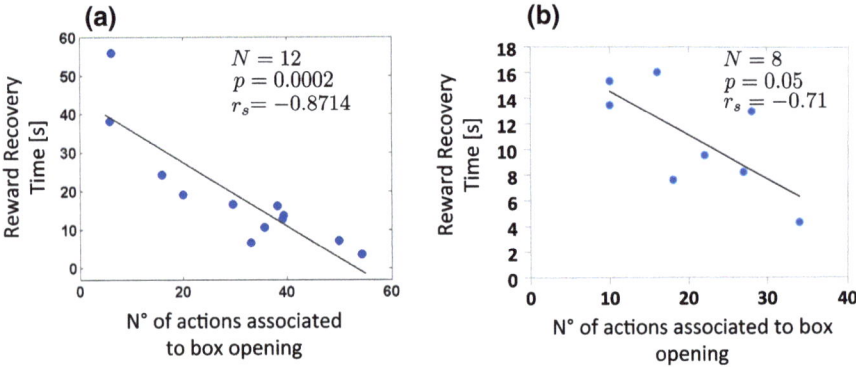

**Fig. 8** Relation between time to retrieve the reward in test phase and levels of object (buttons/module) manipulation performed during free manipulation phase in children (**a**) and in capuchin monkeys (**b**). Correlation test values and relative p levels are reported in each figure

the test phase. By contrast, no correlation was found in CTRL subjects either for children or for monkeys.

*Spontaneous manipulation leads to a better/faster recall of information*

Given the role of action–outcome contingencies in affecting the level of exploration of the board in both studies, we predicted for EXP subjects to outperform CTRLs in the test phase; in this phase, subjects were asked to recall the correct action that would allow the rewarded box to open so as to retrieve the reward inside it. As expected, the performance of EXP children in the test phase resulted to be overall better than that of CRTLs. Although EXP and CTRL children subjects retrieved the same number of rewards, EXP subjects performed fewer actions and needed less time compared to their paired CTRLs (Fig. 9a). Similarly, EXP monkeys obtained a higher percentage of rewards than CTRL subjects and did so in shorter time (Fig. 9b).

*The role of exploration in determining action learning and spatial relation contingencies*

Given the experimental design, during the free manipulation phase, only EXP subjects could acquire knowledge about the functioning of the board. During such phase, two aspects of action selection could be assessed: (i) action learning, i.e., the understanding that an action (extended push in the children study, rotating one of the two modules' tap in the monkey study) causes the opening of the box and

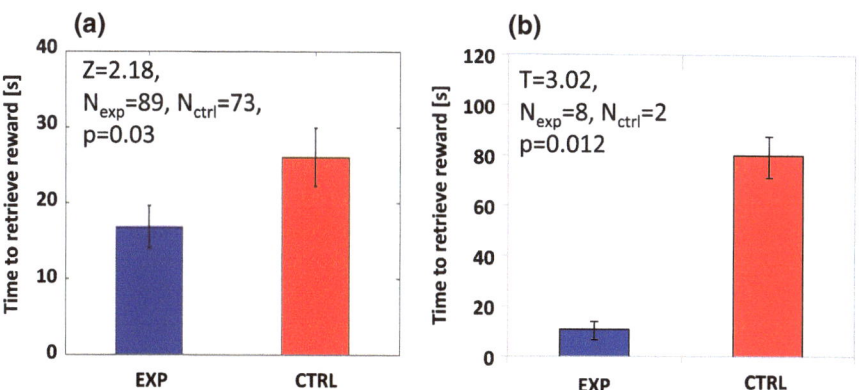

**Fig. 9** Time to retrieve reward in test phase for Exp and CTRL group in both children (**a**) and capuchin monkeys (**b**). Statistical tests of the null hypothesis and their relative p values are reported in each figure

(ii) spatial learning, i.e., the understanding of the correct relation between the object (button/module) and the box it controls. If a subject has learned that an action opens the box, she/he is expected to perform it more frequently when asked to retrieve a reward from inside it. Similarly, if she/he has understood the spatial relation between the object (button/module) and box(es), the correct object should be manipulated more frequently than the others. The results of the children study suggest that EXP subjects refined the spatial relations between buttons and boxes during the test trials and not during the free manipulation phase. CTRL subjects, who were not allowed to learn which action caused the opening of the box during the free manipulation phase, did not have enough time to acquire the spatial relation during the nine trials of the test phase. These results are somewhat similar to those found in the monkey study. Specifically, in the free manipulation phase we found that EXP monkey subjects lacked a preference for the rotation of the correct module over the rotation of the wrong one (e.g., the module not associated with the opening of the box). This finding seems to suggest that in monkeys the association between action and outcome might be established before the association between location and outcome. Given the subtle differences in the experimental setup between the two studies (only one central box in the monkey study vs multiple boxes spatially associated with different buttons in the children study), we cannot further expand this hypothesis, but the fact that EXP children (but not CTRLs) extended their learning of spatial relation during the subsequent test phase seems to corroborate our interpretations.

## 4 Conclusions

Thanks to the primates' propensity to explore objects, in this study we were able to assess how much sensorimotor contingencies due to manipulation can affect the way in which individuals acquire and reuse new knowledge. To this purpose, we compared the explorative behavior of 3- to 4-year-old children and adult capuchin monkeys, during the interaction with a mechatronic board. The board is designed to promote object manipulation through pushing or lifting motions of the hand as a whole or of the digits individually.

Our results suggest that action–outcome associations during spontaneous exploration promotes and maintains subjects' interest toward the board, while the lack of a congruent contingency in the control condition led to a progressive decrease in performance in both children and monkeys. The knowledge acquired in the free manipulation phase from EXP subjects is successfully recalled during the test phase. Indeed, both children and monkeys seem to acquire competence about the functioning of the objects in the free manipulation phase, while EXP children (but not CTRLs) extended their learning of spatial relations during the test phase.

These findings support the hypothesis that sensorimotor contingencies based on manual exploratory behavior play a fundamental role in promoting skill acquisition (even in the absence of extrinsic rewards), and that similar mechanisms operate in human and nonhuman primates. Thus, on the one hand, our results show that the ability to recruit knowledge gained by action–outcome contingencies during free manual exploration of objects develops in the early years of childhood; on the other hand, our data provide insights concerning the evolutionary roots of the mechanisms underlying sensorimotor contingencies. Since New World monkeys (Platyrrhines) in the primate lineage diverged 35–40 million years ago, the fact that these abilities are within the ken of a living primate species which is phylogenetically very distant from humans supports the idea that these skills were already present in the earliest monkeys.

In human and nonhuman primates, the hands are the main effector organs used to explore nearby objects, and the use of mechatronic devices has many potential fields of application in the study of how these species use their hands to acquire knowledge by manipulation activities.

**Acknowledgements** This work was funded by FP7-ICT program (project no. ICT-2007.3.2-231722—IM-CLeVeR)

## References

Aversi-Ferreira, T. A., Maior, R. S., Carneiro-e Silva, F. O., Aversi-Ferreira, R. A., Tavares, M. C., Nishijo, H., & Tomaz, C. (2011). Comparative anatomical analyses of the forearm muscles of *Cebus libidinosus* (Rylands et al. 2000): Manipulatory behavior and tool use. *PloS one, 6*(7), e22165.
Baillie, J. C. (2016). Why alphago is not AI, March 2016. Available at http://spectrum.ieee.org/automaton/robotics/artificial-intelligence/why-alphago-is-not-ai. Cited 20 December 2016.
Bertenthal, B., & Von Hofsten, C. (1998). Eye, head and trunk control: the foundation for manual development. *Neuroscience and Biobehavioral Reviews, 22*(4), 515–520.
Bortoff, G. A., & Strick, P. L. (1993). Corticospinal terminations in two new-world primates: Further evidence that corticomotoneuronal connections provide part of the neural substrate for manual dexterity. *The Journal of Neuroscience, 13*(12), 5105–5118.
Byrne, G., & Suomi, S. J. (1996). Individual differences in object manipulation in a colony of tufted capuchins. *Journal of Human Evolution, 31*(3), 259–267.
Christel, M. I., & Fragaszy, D. (2000). Manual function in *Cebus apella* digital mobility, preshaping, and endurance in repetitive grasping. *International Journal of Primatology, 21*(4), 697–719.
Cioni, G., & Giuseppina, S. (2013). Pediatric neurology part I: Chapter 1. In *Normal psychomotor development. Elsevier Inc. Chapters*, 111.
Costello, M. B., & Fragaszy, D. M. (1988). Prehension in *Cebus* and *Saimiri*: I. grip type and hand preference. *American Journal of Primatology, 15*(3), 235–245.
Cutkosky, M. R. (1989). On grasp choice, grasp models, and the design of hands for manufacturing tasks. *Robotics and Automation, IEEE Transactions on, 5*(3), 269–279.
Fallang, B., Saugstad, O. D., & Hadders-Algra, M. (2000). Goal directed reaching and postural control in supine position in healthy infants. *Behavioural Brain Research, 115*(1), 9–18.
Focaroli, V., Taffoni, F., & Iverson, J. M. (2015). Motor planning ability in typically developing children and children with autism spectrum disorder. *Psicologia Clinica dello Sviluppo, 19*(1), 3–26.
Fragaszy, D. M., & Adams-Curtis, L. E. (1991). Generative aspects of manipulation in tufted capuchin monkeys (*Cebus apella*). *Journal of Comparative Psychology, 105*(4), 387.

Fragaszy, D. M., & Boinski, S. (1995). Patterns of individual diet choice and efficiency of foraging in wedge-capped capuchin monkeys (*Cebus olivaceus*). *Journal of Comparative Psychology, 109*(4), 339.

Fragaszy, D. M., & Crast, J. (2016). Functions of the hand in Primates. In *The evolution of the primate hand: Anatomical, developmental, functional, and paleontological evidence* (pp. 313–344). New York: Springer.

Fragaszy, D. M., Visalberghi, E., & Fedigan, L. M. (2004). *The complete capuchin: The biology of the genus Cebus*. Cambridge: Cambridge University Press.

Iverson, J. M. (2010). Developing language in a developing body: The relationship between motor development and language development. *Journal of Child Language, 37*(02), 229–261.

Kapandji, I. A. (1987). *The physiology of the joints: Lower limb* (vol. 2). Elsevier Health Sciences.

Kaplan, F., & Pierre-Yves, O. (2007). In search of the neural circuits of intrinsic motivation. *Frontiers in Neuroscience, 1*(1), 225–242.

Lynch Alfaro, J. W., Boubli, J. P., Olson, L. E., Di Fiore, A., Wilson, B., Gutiérrez-Espeleta, G. A., et al. (2012a). Explosive pleistocene range expansion leads to widespread amazonian sympatry between robust and gracile capuchin monkeys. *Journal of Biogeography, 39*(2), 272–288.

Lynch Alfaro, J. W., Silva, J. D. S. E., & Rylands, A. B. (2012b). How different are robust and gracile capuchin monkeys? An argument for the use of *Sapajus* and *Cebus*. *American Journal of Primatology, 74*(4), 273–286.

Napier, J. R. (1956). The prehensile movements of the human hand. *Bone & Joint Journal, 38*(4), 902–913.

O'Regan, J. K., & Noë, A. (2001). A sensorimotor account of vision and visual consciousness. *Behavioral and Brain Sciences, 24*(05), 939–973.

Out, L., van Soest, A. J., Savelsbergh, G. J., & Hopkins, B. (1998). The effect of posture on early reaching movements. *Journal of Motor Behavior, 30*(3), 260–272.

Padberg, J., Franca, J. G., Cooke, D. F., Soares, J. G., Rosa, M. G., Fiorani, M., et al. (2007). Parallel evolution of cortical areas involved in skilled hand use. *The Journal of Neuroscience, 27*(38), 10106–10115.

Panger, M. A. (1998). Object-use in free-ranging white-faced capuchins (*Cebus capucinus*) in costa rica. *American Journal of Physical Anthropology, 106*(3), 311–321.

Perry, S., & Manson, J. H. (2008). *Manipulative monkeys*. Harvard University Press.

Polizzi di Sorrentino, E., Sabbatini, G., Truppa, V., Bordonali, A., Taffoni, F., Formica, D., et al. (2014). Exploration and learning in capuchin monkeys (*Sapajus* spp.): the role of action–outcome contingencies. *Animal Cognition, 17*(5), 1081–1088.

Preuschoft, H., & Chivers, D. J. (2012). *Hands of primates*. Berlin: Springer.

Rochat, P., & Goubet, N. (1995). Development of sitting and reaching in 5-to 6-month-old infants. *Infant Behavior and Development, 18*(1), 53–68.

Sabbatini, G., Meglio, G., & Truppa, V. (2016). Motor planning in different grasping tasks by capuchin monkeys (*Sapajus* spp.). *Behavioural Brain Research, 312*, 201–211.

Spencer, J. P., Vereijken, B., Diedrich, F. J., & Thelen, E. (2000). Posture and the emergence of manual skills. *Developmental Science, 3*(2), 216–233.

Spinozzi, G., Castorina, M. G., & Truppa, V. (1998). Hand preferences in unimanual and coordinated-bimanual tasks by tufted capuchin monkeys (*Cebus apella*). *Journal of Comparative Psychology, 112*(2), 183.

Spinozzi, G., Laganà, T., & Truppa, V. (2007). Hand use by tufted capuchins (*Cebus apella*) to extract a small food item from a tube: Digit movements, hand preference, and performance. *American Journal of Primatology, 69*(3), 336–352.

Spinozzi, G., Truppa, V., & Laganà, T. (2004). Grasping behavior in tufted capuchin monkeys (*Cebus apella*): Grip types and manual laterality for picking up a small food item. *American Journal of Physical Anthropology, 125*(1), 30–41.

Taffoni, F., Formica, D., Schiavone, G., Scorcia, M., Tomassetti, A., Polizzi di Sorrentino, E., et al. (2013). The "mechatronic board": A tool to study intrinsic motivations in humans, monkeys, and humanoid robots. In *Intrinsically motivated learning in natural and artificial systems* (pp. 411–432). Berlin: Springer.

Taffoni, F., Formica, D., Zompanti, A., Mirolli, M., Balsassarre, G., Keller, F., & Guglielmelli, E. (2012a). A mechatronic platform for behavioral studies on infants. In *2012 4th IEEE RAS & EMBS International Conference on Biomedical Robotics and Biomechatronics (BioRob)* (pp. 1874–1878). IEEE.

Taffoni, F., Tamilia, E., Focaroli, V., Formica, D., Ricci, L., Di Pino, G., et al. (2014). Development of goal-directed action selection guided by intrinsic motivations: an experiment with children. *Experimental Brain Research, 232*(7), 2167–2177.

Taffoni, F., Vespignani, M., Formica, D., Cavallo, G., Polizzi di Sorrentino, E., Sabbatini, G., et al. (2012b). A mechatronic platform for behavioral analysis on nonhuman primates. *Journal of Integrative Neuroscience, 11*(01), 87–101.

Terborgh, J. (1983). *Five New World primates: A study in comparative ecology*. Princeton University Press.

Truppa, V., Spinozzi, G., Laganà, T., Mortari, E. P., & Sabbatini, G. (2016). Versatile grasping ability in power-grip actions by tufted capuchin monkeys (*Sapajus* spp.). *American Journal of Physical Anthropology, 159*(1), 63–72.

Visalberghi, E. (1988). Responsiveness to objects in two social groups of tufted capuchin monkeys (*Cebus apella*). *American Journal of Primatology, 15*(4), 349–360.

Visalberghi, E., Cavallero, S., Fragaszy, D. M., Izar, P., Aguiar, R. M., & Truppa, V. (2015a). Making use of capuchins' behavioral propensities to obtain hair samples for DNA analyses. *Neotropical Primates, 22*, 89–93.

Visalberghi, E., & Fragaszy, D. (2006). What is challenging about tool use? the capuchin's perspective. In E. A. Wasserman, T. R. Zentall (Eds.), *Comparative cognition: Experimental explorations of animal intelligence* (pp. 529–552).

Visalberghi, E., Sirianni, G., Fragaszy, D., & Boesch, C. (2015b). Percussive tool use by Taï western chimpanzees and fazenda Boa Vista bearded capuchin monkeys: A comparison. *Philosophical Transactions R Society B, 370*(1682), 20140351.

Zander, S. L., & Judge, P. G. (2015). Brown capuchin monkeys (*Sapajus apella*) plan their movements on a grasping task. *Journal of Comparative Psychology, 129*(2), 181.

Zander, S. L., Weiss, D. J., & Judge, P. G. (2013). The interface between morphology and action planning: A comparison of two species of New World monkeys. *Animal Behaviour, 86*(6), 1251–1258.

# Current Achievements and Future Directions of Hand Prostheses Controlled via Peripheral Nervous System

Anna Lisa Ciancio, Francesca Cordella, Klaus-Peter Hoffmann, Andreas Schneider, Eugenio Guglielmelli and Loredana Zollo

**Abstract** The human hand is a powerful tool to feel and act on the environment and a very sophisticated means for physical and social interaction. This is why hand loss can be perceived as a devastating damage that changes people lifestyle. It causes a severe impairment for the amputees and can significantly alter their quality of life, since it affects personal and working fields by reducing the level of autonomy, the capability of performing activities of daily living (ADLs), and the capability to gesture and interact with other people. The upper-limb amputation involves almost 4000 people per year in Italy and about the 20% of amputations in USA. The relevance of the upper-limb loss in the international scenario motivates the flourishing research in the field of upper-limb prosthetics. This chapter intends to provide an overview on hand prostheses driven by non-invasive and invasive interfaces with the peripheral nervous system (PNS), taking into account technical aspects related to hand control, peripheral interfaces, and clinical features about the restoration of sensory feedback. The international scenario of off-the-shelf and on-the-shelf prosthetic hands is explored, and pros and cons of technologies are analyzed. This chapter is especially focused on the recent studies on the restoration

---

A.L. Ciancio (✉) · F. Cordella · E. Guglielmelli · L. Zollo
Department of Engineering, Università Campus Bio-Medico di Roma, Rome 00128, Italy
e-mail: a.ciancio@unicampus.it

F. Cordella
e-mail: f.cordella@unicampus.it

E. Guglielmelli
e-mail: e.gugliemelli@unicampus.it

L. Zollo
e-mail: l.zollo@unicampus.it

K.-P. Hoffmann · A. Schneider
Fraunhofer Institut für Biomedizinische Technik, 66386 St. Ingbert, Germany
e-mail: Klaus-Peter.Hoffmann@ibmt.fraunhofer.de

A. Schneider
e-mail: Andreas.Schneider@ibmt.fraunhofer.de

© Springer International Publishing AG 2017
M. Bertolaso and N. Di Stefano (eds.), *The Hand*, Studies in Applied Philosophy, Epistemology and Rational Ethics 38, DOI 10.1007/978-3-319-66881-9_5

of tactile perception in amputees through neural interfaces and first evidence on bidirectional hand control. Current achievements on this thorny topic are in-depth explained in this chapter and future directions are finally roughed out.

**Keywords** PNS-based prosthetic hand · Upper-limb prosthesis · Sensory feedback · Grasping · Manipulation

## 1 Introduction

The human hand is a powerful tool characterized by a complex mechanical structure and a sophisticated sensory system that enables dexterous grasping and manipulation tasks through a bidirectional communication with the brain. The hand loss causes the interruption of this communication and, consequently, severe impairments for the subjects may arise, regarding both motor control and sensory feedback. The subject's quality of life is altered for the personal as well as the working spheres; the capability to perform activities of daily living (ADLs) and interact with other people is significantly affected.

Worldwide, an estimated three million of people suffer from an arm or hand amputation (LeBlanc 2008). Approximately 3500 and 5200 upper-limb amputations are reported each year in Italy and in UK, respectively. The different levels of upper-limb loss have the following incidence on the total amputations: 16% transhumeral, 12% transradial, 2% forequarter, 3% shoulder disarticulation, 1% elbow disarticulation, 2% wrist disarticulation, 61% transcarpal, and 3% bilateral limb loss.

In the last 70 years, the advances in technological and surgical field have produced significant improvements in the upper-limb prosthetics. Hand design, control, and sensory feedback have been fostered to realize prostheses able to reproduce aesthetical as well as functional features of the lost limb, in order to meet prosthetic user needs (Cordella et al. 2016). However, despite the advances, today's upper-limb prostheses are still affected by relevant limitations: lack of an intuitive and reliable interface, lack of sensory feedback, noise produced by the actuators.

This chapter intends to critically review the state of the art on hand prostheses, by reporting main requirements coming from the analysis of users' needs and describing main recent technical and clinical achievements. This chapter is especially focused on the technologies enabling the restoration of tactile perception and bidirectional control in amputees and reports the latest human studies on hand control via neural interfaces.

This chapter is structured as follows: In Sect. 2, an overview of the users' needs is provided and main requirements for upper-limb prostheses are reported. Section 3 discusses off-the-shelf and on-the-shelf prosthetic hands and myoelectric control. In Sect. 4, peripheral neural interfaces for prosthesis control and afferent stimulation are introduced and compared. Section 5 provides an overview of recent human experiments on the restoration tactile sensing and bidirectional hand control. Finally, Sect. 6 draws the conclusions.

## 2 User Needs

Over the years, many attempts have been made to provide the amputees with a valid prosthetic substitute for the lost limb. Therefore, several studies have been devoted to identify (i) the type of activities that the prostheses can help perform (Van Lunteren et al. 1983); (ii) the reason why several amputees prefer not to use the prosthesis (Biddiss and Chau 2007; Peerdeman et al. 2011); and (iii) job-related problems before and after amputation (Wright et al. 1995). Table 1 summarizes main studies on user needs, accounting for user experience with myoelectric,

**Table 1** Main studies focused on the analysis of user needs, considered population (in terms of type of prosthesis and level of limb loss), main questions shared among the studies and corresponding answers

| Study | Type of prosthesis | Level of limb loss | Questions | Answers |
|---|---|---|---|---|
| Kyberd and Hill (2007) | 60% C, 27% Myo, 13% other | 58% Tr, 31% Th, 7% Sd | 1 | More natural appearance Improvements in movement and grip functions |
| Biddiss and Chau (2007) | BP, C, E | 54% Tr, 21% Th, 7% Sd, 16% Wd, 15% Bi | 1 | Comfort, Function, Comfort |
| | | | 2 | Household maintenance Cooking, eating, dressing, personal hygiene, typing |
| Jang et al. (2011) | 80.2% C, 1% Myo, 79.2% other | 6.6%Sd, 20.5%Th, 48.4%Tr, 6.6%Wd, 17.9%Tc, 11% Bi | 1 | Cosmesis and comfort |
| | | | 2 | Cooking, eating, dressing, personal hygiene, typing |
| Pylatiuk et al. (2007) | Myo | 76.9%Tr, 14.8%Th, 5.5% not specified | 1 | Sensory feedback |
| | | | 2 | Using cutlery |
| Østlie et al. (2012) | 19.9% C, 34.2% Myo, 29.8% BP, 16,1% other | 85% Tr, 15% Th | 2 | Cooking, eating, dressing, personal hygiene |
| Østlie et al. (x2012) | 7% BP, 8% Myo, 25% both | 71.2% Tr, 28.8% Sd and Th, 4% Bi | 2 | Eating, personal hygiene, employment and recreation |
| Lucchetti et al. (2015) | Myo | Tr | 1 | Functionality |
| | | | 2 | Eating and dressing |

Legend:
*Tr* Transradial, *Th* Transhumeral, *Sd* Shoulder disarticulation, *Tc* Transcarpal, *Bi* Bilateral, *Wd* Wrist disarticulation, *BP* Body-powered, *E* Electric, *C* Cosmetic, *Myo* Myoelectric
Question 1 = Consumer design priorities
Question 2 = ADLs the subjects would like to perform

electric, body-powered, and passive upper-limb prostheses. Notwithstanding the high variability of the users' answers, a common set of needs and requirements can be identified (Cordella et al. 2016). They are listed below:

- To perform activities of daily living mainly related to eating, dressing, type writing, handling a cell phone, and opening the door (Kyberd and Hill 2007; Biddiss et al. 2007; Cloutier and Yang 2013). This entails that the prosthetic system needs to perform basic grasping actions (i.e., power, pinch, lateral, neutral, and pointing of the index finger) and simple manipulation tasks enabling the execution of activities of daily living (ADLs).
- To feel what is grasped or manipulated through sensory feedback.
- To perform actions in a more coordinated manner and with less visual attention. In particular, the use of a control system able to manage position and force exerted by the fingers on the objects can lighten the role of the visual feedback giving more importance to sensory feedback.
- To perform actions requiring fine force control.
- To change position and orientation of the grasped object, thus entailing capabilities of object manipulation; to move each finger independently, as in free manipulation; to improve the performance of thumb, index, and middle finger, in order to increase precision and efficient handling of small objects; and to provide the prosthetic hand with a wrist module, given the fundamental role played in ADLs.
- To wear prostheses with high level of anthropomorphism (in terms of size, weight, shape, color, and achievable grasping configurations) and with low motor noise.

Most of the required grasping and manipulation capabilities depend on the technical features of the multifingered prostheses and on the implemented control strategies that notably affect functionality.

Although several attempts have been done to provide the user with an intuitive and effective prosthesis control, the so far proposed solutions are not able to manage and properly combine basic hand movements to generate the desired complex motion. The literature is especially focused on intuitive control approaches, where the user's intention is extracted from peripheral signals through pattern recognition techniques (see Sect. 3).

On the other hand, upper-limb prosthesis users also express their necessity to feel and interact with the world through the prosthesis. This requires to provide the prosthetic hand with a tactile sensory system for the twofold purpose of performing a force control during grasping and returning force/tactile sensation to the user by means of peripheral interfaces on the afferent pathway. Over the years, different approaches have been proposed for eliciting tactile sensations (Antfolk et al. 2013; Schofield et al. 2014) as it will be detailed in Sect. 5.

## 3 Hand Prostheses and Myoelectric Control

Upper-limb prostheses can be classified into two main categories based on their functioning: passive and active prostheses (Fumero and Costantino 2001). Passive prostheses are in turn divided into cosmetic and functional ones; active prostheses include body-powered and externally powered prostheses, which are further classified into myoelectric and electric.

The most advanced commercially available myoelectric multifingered prosthetic hands are Touch Bionics i-Limb,[1] Ottobock Michelangelo[2] and RSL Steeper BeBionic[3] (Fig. 1). Table 2 summarizes the main characteristics of multifingered commercial and research hands. Further details about mechanical characteristics of anthropomorphic poliarticulated prostheses are available in Belter et al. (2013).

The commercial hands are able to provide different grip patterns but they are still characterized by a limited number of active Degrees of Freedom (DoFs) (5 at most) and do not provide the user with sensory feedback (Table 2). The research hands can guarantee a major number of active DoFs, applying lower force level respect to commercial prostheses. In both cases, the noise produced by the actuators during movements makes the prosthetic hands still far from fully addressing the users' needs (Clement et al. 2011). Moreover, all the myoelectric prostheses only use a position loop to control hand grasping, forcing the user to continuously look at the object for trying to regulate the grasping force, also for preventing object slippage. Therefore, the user requirement to control the grasping providing less visual attention is not fulfilled.

Providing a prosthetic device with reliable tactile information still represents a challenge in the robotic and prosthetic fields.

The subject intention of movement is extracted from muscular signals through EMG interfaces.

The on/off control is a simple and intuitive control modality allowing the activation of a defined prosthesis function when the EMG signals exceed a threshold. This modality requires many sites to extract the EMG signal, one for each function to control.

The agonist/antagonist myoelectric control (Popov 1965) is the most adopted solution for commercially available myoelectric prostheses, thanks to its simplicity and robustness (Jiang and Farina 2014). A couple of electrodes on agonist/antagonist muscles allows to associate to the contraction of one muscle the motion of opening and to the contraction of the other muscle the motion of closure, both with a constant speed (Popov 1965). The simultaneous contraction of both muscles permits to switch between different functions.

Proportional control (Fougner et al. 2012) allows to vary force and speed proportionally to the amplitude of the EMG signals recorded from a pair of

---

[1]http://www.touchbionics.com/.
[2]http://www.living-with-michelangelo.com/home/<.
[3]http://www.bebionic.com.

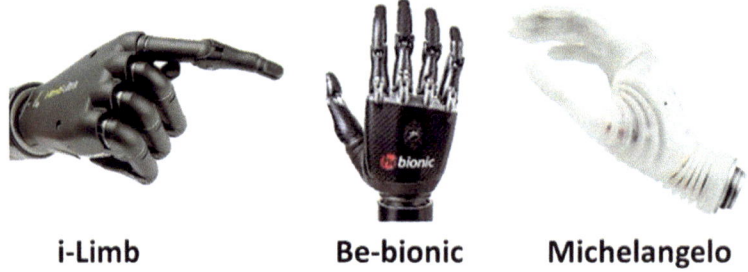

**Fig. 1** Most advanced commercially available prosthetic hands

agonist/antagonist muscles. Hence, the voltage command for the motors is taken as proportional to the contraction intensity. Muscles co-contraction allows to select the degree of freedom to control.

In this control, only a limited number of DoFs can be independently controlled (far from the multifunctional control of the human hand) (Popov 1965). In particular, it is possible to perform a limited number of hand configurations, but it is not possible to move each finger independently (as it is desired by the users).

The commercially available myoelectric prostheses use classical myoelectric control with on/off or agonist/antagonist EMG control for selecting the DoF to be controlled.

In order to overcome these limitations, several alternative approaches can be found in the literature, such as ultrasound imaging (Gonzales and Castellini 2013), force myography (FMG) (Wininger et al. 2008), and, most importantly, pattern recognition techniques (Cloutier and Yang 2013) applied to EMG signals acquired through implantable (IMES Pasquina et al. 2015) or surface electrodes (Dohnalek et al. 2013). Most of them are still used only in the research field. Pattern recognition consists of the following steps (Fig. 2): (a) feature extraction in time or frequency domain (Cloutier and Yang 2013); (b) dimensionality reduction; and (c) classification. Pattern recognition classifiers can be grouped into linear classifiers, such as linear discriminant analysis (LDA) or perceptron or support vector machine (SVM), nonlinear classifiers, such as nonlinear logistic regression or SVM with nonlinear kernels, and multilayer perceptron or multilayer SVM (Ortiz-Catalan et al. 2014; Ciancio et al. 2016). Performance is affected by arm posture modifications, the complex nature of forearm muscles synergies, inherent cross talk in the surface signal, and displacement of the muscles during contraction. Furthermore, performance in the real context seems different from the laboratory settings (Li et al. 2012), thus limiting the clinical applicability of pattern recognition approach. The first commercial device based on pattern recognition and surface electrodes is the COAPT,[4] which appeared on the market is in January 2015.

---

[4]COAPT Complete control, 2014. Available: http://www.coaptengineering.com/.

Table 2 Characteristics of poliarticulated commercially available prosthetic hands (i-Limb, Bebionic and Michelangelo) and of research hands with application in prosthetics (Southampton, UB hand III, IH2 Azzurra) (Belter et al. 2013)

| Hand and company name | i-Limb by Touch Bionics | Bebionic by RSL Steeper | Michelangelo by Ottobock | Southampton hand | UB hand III | IH2 Azzurra by Prensilia |
|---|---|---|---|---|---|---|
| Weight (g) | 443–515 | 550–598 | 420 | 400 | – | 640 |
| No of actuators | 5 DC motors | 5 DC motors | 2 DC motors | 6 DC Motors | 20 DC motors | 5 DC motors |
| Active DoFs | F/E of MCP joint of each finger and thumb opposition | F/E of MCP joint of each finger | F/E of all the fingers contemporarily and thumb opposition | F/E of MCP joint of each finger and thumb opposition | F/E of MCP, PIP and DIP of thumb, index abduction, and middle, index abthumb opposition, F/E of MCP and PIP of ring and little. | F/E of MCP joint of thumb, index, middle, ring (coupled with little) and thumb opposition |
| Joint coupling mechanism | Tendon linking MCP to PIP | Linkage spanning MCP to PIP | Cam design with links to all fingers | Worm wheels gears | Tendon driven mechanism | Tendon driven mechanism |
| Grasping configuration | Power, Precision, Lateral, Hook, Finger-point | Power, Precision, Lateral, hook, finger-point | Opposition, Lateral, Neutral Mode | Power, Precision, Lateral, Hook | Power, Precision, Lateral, Hook, Finger-point | Power, Precision, Lateral, Hook, Finger-point |
| Maximum applied force | 100–136 N | 140 N | 70 N | Fingertip forces: 9 N | Tendon force: 70 N | 35 N |

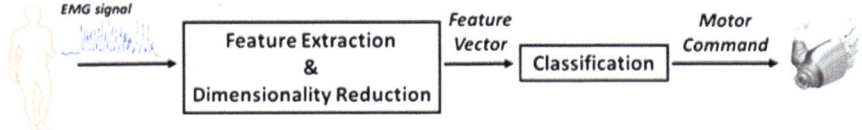

**Fig. 2** Block scheme of a myoelectric hand control

In order to provide intuitive and stable myoelectric control, the adoption of IMES has been tested on transradial amputees (Pasquina et al. 2015) with very promising preliminary results (especially for robustness to limb position and environmental conditions). However, IMES cannot be employed when the sensing sites are very close each other, or when the target muscle is small or thin.

Notwithstanding the huge amount of research on myoelectric control, great effort is still required for improving hand control and interfaces, and move toward clinical translation of research results. Currently, active research challenges are related to (i) real-time, direct, robust, and simultaneous control of multiple DoFs in a natural and intuitive manner, (ii) bidirectional communication with the peripheral nervous system (PNS), and (iii) fast learning of hand control.

## 4 Neural Interfaces for Prosthesis Control and Afferent Stimulation

Electrodes can be used as interface between a technical system and biological tissue to record bioelectrical potentials or to stimulate electrically neuronal structures. They are transducers to transform current: ionic into electric or electric into ionic. Electrode design and material selection are based on factors such as access to and location of the application site and the application duration. The selectivity depends on the invasiveness and the size of electrode contacts (Navarro et al. 2005). Stiff and flexible electrode structures are used.

Structural or backbone materials of the electrodes are often polyimide, silicone, or silicon. Advantages of polyimide structures are the high flexibility and thinness (10–15 μm), the low density (1.42 g/cm$^3$), the water uptake (<0.5%), and the good biostability (>12 months). Electrode contact materials, the electrically active part of the electrode, are metals (e.g., gold, platinum), metal compounds (e.g., titanium nitride, iridium oxide), or intrinsically conductive polymers (e.g., PEDOT). All materials have to be biocompatible and longtime biostable. Stimulation electrodes should have a high charge injection capacity to transfer charge without causing chemical reactions. The stimulation has to be charge-balanced. Thereby the influence of redox reactions and corrosions is reduced, and the tissue can be protected (Neural Stimulation and 2008).

To evaluate the microfabricated electrodes, electrochemical, mechanical, optical, and biological characterizations are necessary. Preferred tests in addition to validation of biocompatibility are impedance spectroscopy as well as determination of charge injection capacity using pulse tests.

The final step is sterilization before the electrodes can be used in in vivo studies. ethylene oxide (ETO) sterilization allows the treatment of electrodes that are already packaged for shipment.

Bionic hand prostheses have to restore both motor and sensory functionalities. Therefore, the neural interface should achieve a bidirectional communication with afferent and efferent nerve fibers. The required selectivity could be realized with implantable microelectrodes. They are very invasive and are placed around (circumneural, e.g., cuff electrode, FINE), within (interfascicular e.g., shaft electrode or intrafascicular LIFE, TIMES, ds-FILE), and between (intraneural, e.g., sieve electrode) the nerves. Invasiveness and selectivity increase in referred order. They acquire bioelectrical activity from motor nerve fibers to control the hand prostheses as well as to stimulate the sensor fibers to realize a sensory feedback.

The feasibility with tf-LIFE (thin film—Longitudinal Intrafascicular Electrode) was shown by Rossini et al. (2010). The electrodes were placed in median nerves and ulnar nerves of an amputee. Individual fingers of the hand prostheses could be selectively moved. The electrical nerve stimulation evoked sensations. The sensory feedback with TIME (transverse intrafascicular multichannel electrode) allowed the discrimination between the shape and texture of different objects (Raspopovic et al. 2015). A comparative analysis of CUFF, tf-LIFE, and TIME shows that the threshold for muscle activation with TIME and tf-LIFE is significantly lower than with CUFF. The selectivity to activate muscles is higher for TIME compared with tf-LIFE (Badia et al. 2011). To increase the selectivity of CUFF electrodes and to decrease the stimulation thresholds, the shape of this electrode structure is flatten. The FINE (flat interface nerve electrode) is described by Tyler and Durand (2002).

A future-oriented development is the ds-FILE (double-sided filament electrode) (Poppendieck et al. 2015). In Figure 3), the design and details of the ds-FILE are shown. Table 3 gives an overview about three different intrafascicular electrode structures.

The papers (Navarro et al. 2007) and (Ciancio et al. 2016) give a good overview and critical review on interfaces with the PNS to control prosthetic hands.

Electrical nerve stimulation is a proven method to activate peripheral nerves. Sometimes, the electrode contacts are not in the optimal position for stimulation or recording. Microactuators embedded in the flexible electrode structures on the basis of shape memory alloys could change the electrode contact position and could improve the electrical connection to the nerve tissue (Bossi et al. 2007).

The use of electrodes includes also disadvantages. The invasive access and the positioning directly on or in the nerve depend on mechanical nerve manipulation. tf-LIFE, TIME, and ds-FILE penetrate the nerve. The electrode and electrical stimulation could change the ambient tissue. This could decrease the longtime stability. So, new stimulation techniques such as ultrasound, optogenetic, or biochemical

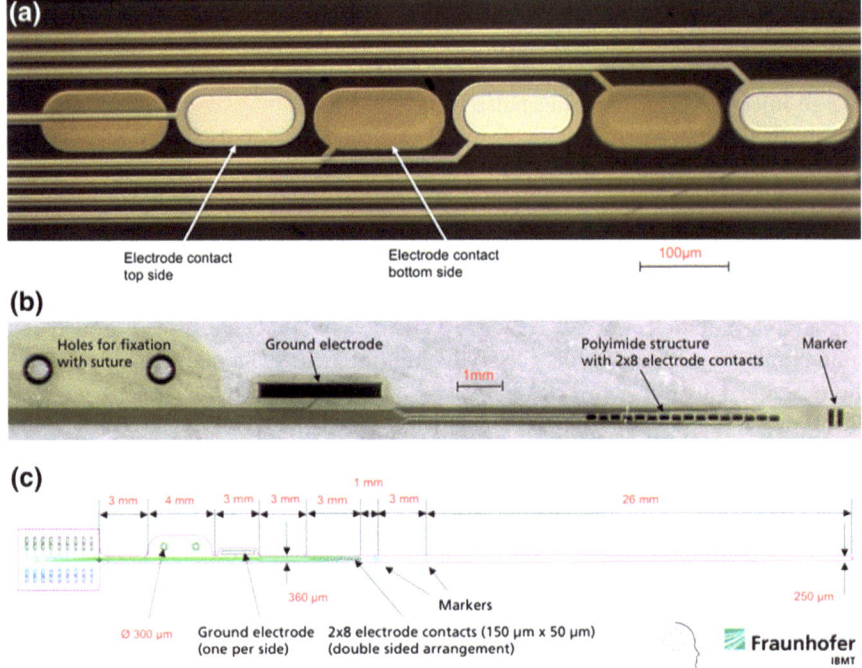

**Fig. 3** Double-sided filament electrode (ds-FILE) for stimulation and re-cording **a** detail **b** polyimide structure **c** design

stimulation have to be developed to achieve a longtime stable stimulation. This will be an important step in restoring the sensor functionality of hand prosthesis.

## 5 Tactile Feedback and Neural Control of Upper-Limb Prostheses

A closed-loop control around the user (Antfolk et al. 2013) is characterized by a bidirectional communication between the user (mainly the PNS) and the prosthetic system (Fig. 4) via the following modules:

(1) The interface combines recording and stimulation capabilities and is responsible for the communication with the PNS through the efferent and afferent pathways (Sect. 4).
(2) The control system drives the prosthesis actuators on the basis of proprioceptive and tactile/force sensory information. User intention is decoded by the signals recorded on the efferent pathway by means of EMG or ENG interfaces and used to generate the control commands (Sect. 3).

**Table 3** Comparison of different intrafascicular electrodes

|  | tf-LIFE | TIME | ds-FILE |
| --- | --- | --- | --- |
| Type | Thin film—Longitudinal Intrafascicular Electrode | Transverse Intrafascicular Multichannel Electrode | Double-sided filament electrode |
| Form | Loop, electrode contacts single sided | Loop, electrode contacts single sided | Single filament, electrode contacts double sided |
| Materials | Platinum on a sub layer of polyimide | Platinum and iridium oxide on a sub layer of polyimide | Microrough platinum on a sub layer of polyimide |
| Electrode contacts | 8 electrode contacts | 10 electrode contacts | 16 electrode contacts |
|  | 2 ground contacts | 2 ground contacts | 2 ground contacts |
| Size of electrode contacts | 40 μm × 100 μm | Ø 60 μm | 50 μm × 150 μm |
|  | 4.000 μm$^2$ | 2.827 μm$^2$ | 7.500 μm$^2$ |
| Implantation tool | Polyimide loop with needle | Polyimide loop with needle | Surgical needle directly connected to the polyimide filament |
| Application | Recording and stimulation | Recording and stimulation | Recording and stimulation |
| References | Micera et al. (2008), Benvenuto et al. (2010), Hoffmann and Micera (2011) | Boretius et al. (2010), Jensen et al. (2010), Badia et al. (2011) | Poppendieck et al. (2015) |

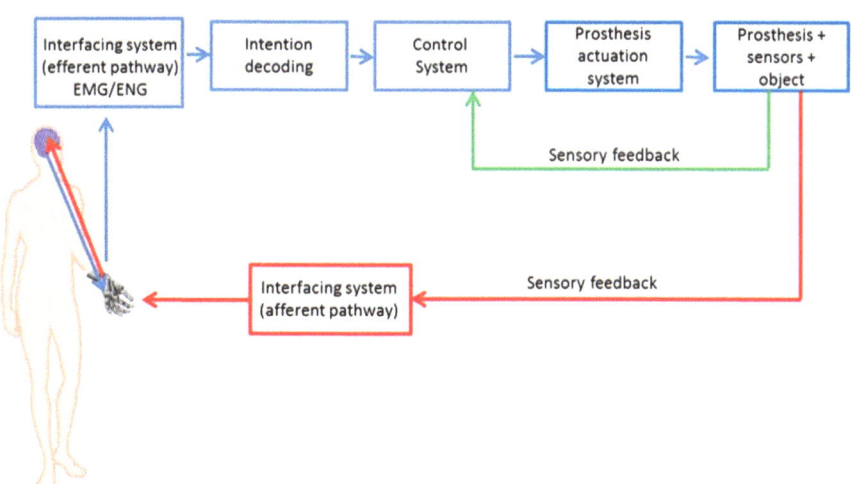

**Fig. 4** Block scheme of the PNS-based control of a prosthetic system

(3) The sensory system returns the tactile feedback about the manipulated object to the control system and, also, to the amputee via the peripheral interface.

The restoration of sensory information regarding the interaction between prosthesis and environment (tactile perception, proprioception, pain, and temperature) (Farina and Aszmann 2014) is carried out through the afferent pathway. In (Childress 1980), three different afferent pathways were proposed, based on (i) visual or auditory feedback signals; (ii) somatic sensory signals, i.e., tactile, proprioception, and vibration; and (iii) feedback signals intrinsic to the prosthesis control system, which use information of the sensors embedded in the prosthesis for automatically adjusting the grasping force.

Somatic sensory signals can be generated through non-invasive or invasive interfacing techniques (Antfolk et al. 2013) (Schofield et al. 2014) such as vibrotactile, electrotactile, mechanotactile, targeted sensory reinnervation (TSR), and neural stimulation. Despite the recent interest in the literature, neural stimulation represents the most promising technique because it allows to exploit the physiological pathways of communication between the hand and the PNS.

Over the years, several studies have been performed in order to investigate the possibility of restoring sensory feedback in individuals with limb loss by means of peripheral nerve stimulation. Clippinger et al. (1974) performed one of the first experiments implanting an induction-powered radio receiver-pulse generator for motor stimulation of the median nerve of 15 patients. This study provided evidence on the possibility of restoring the pressure sensation applied to the grasped object.

In Dhillon and Horch (2005), it has been demonstrated the feasibility of performing a natural control of the prosthesis and returning a natural sensory feedback to the amputee in a closed-loop control by means of implantable peripheral interfaces. Implanted peripheral nerve electrodes were used to (i) elicit touch and movement sensations and (ii) record motor neuron activity usable as hand control signals, without exploring closed-loop, non-visual control of the prosthesis. More recent experimental studies (Rossini et al. 2010; Raspopovic et al. 2015; Ortiz-Catalan et al. 2014; Tan et al. 2014) have also shown that it is possible to restore the natural tactile sensory feedback through peripheral neural interfaces. Table 4 and Fig. 5 provide a brief overview of the aforementioned experimental studies and their results.

The work in Rossini et al. (2010) investigates the feasibility of (i) controlling a hand prosthesis by means of the neural signal directly extracted from median and ulnar nerves of an amputee; (ii) using afferent neural stimulation to elicit tactile sensory feedback. Four tf-LIFE4s (Hoffmann and Kock 2005) electrodes have been implanted, by means of a surgical intervention, two in median and two in ulnar nerves in a parallel way respect to the main nerve axis. Initially, motor output from efferent fibers has been recorded in order to control the prosthetic hand. The classification improved from 75 to 85% in two days. Subsequently, the experimenters stimulated afferent fibers to elicit sensation. The stimulation was artificially triggered by the experimenters without the use of sensors embedded in the prosthesis and the delivering of electrical current in the tf-LIFE4s electrodes allowed to

**Table 4** Summary of results on neural implant studies for sensory feedback restoration

| | Rossini et al. (2010) | Raspopovic et al. (2015) | Ortiz-Catalan et al. (2014) | Tan et al. (2014) |
|---|---|---|---|---|
| Number of subjects | 1 | 1 | 1 | 2 |
| Experimental period | 4 weeks | 4 weeks | Up to 16 months | Up to 24 months |
| Electrodes | tf-LIFEs (thin-film Longitudinally-implanted Intra Fascicular Electrodes) | TIMEs (transversal intrafascicular multichannel electrodes) | Cuff electrode (Ardiem Medical) | FINE (flat interface nerve electrodes) |
| | | | | Cuff electrode (Ardiem Medical) |
| Number of electrodes | 4 | 4 | 1 | Subject 1: 2 FINEs, 1 cuff |
| | | | | Subject 2: 2 FINEs |
| Nerves | Median and ulnar nerves | Median and ulnar nerves | Ulnar nerve | Subject 1: median and ulnar nerves |
| | | | | Subject 2: median and radial nerves |
| Trains of pulses | Rectangular cathodal pulses | Rectangular cathodal pulses | Single active charge-balanced biphasic pulse | Square electrical pulses |
| Frequency | 10–100 Hz | 50 Hz | 8–20 Hz | 10–125 Hz |
| Current | 10-100 µA | Maximum stimulation current: 240 µA (at 100 µs) for the index finger and 160 µA (at 50 µs) for the little finger | 30–50 µA | 1.1–2 mA |
| Pulse width | 10–300 µs | – | – | 24–60 µs |
| Charge | 0.1–4 nC | Median nerve: 14–24 nC | 100-180 µA | Subject 1: 40.7–95.5 nC |
| | | Ulnar nerve: 4–8 nC | | Subject 2: 95–141 nC |
| Elicited hand areas | Figure 6a | Figure 6b | Figure 6c | Figure 6d–e |

(continued)

**Table 4** (continued)

|  | Rossini et al. (2010) | Raspopovic et al. (2015) | Ortiz-Catalan et al. (2014) | Tan et al. (2014) |
|---|---|---|---|---|
| Grasping task | Power grip, pinch grip, little finger flexion | Palmar grasp, pinch grasp, ulnar grasp | Tripod grasp during arm oscillation, power grasp in different limb position | – |

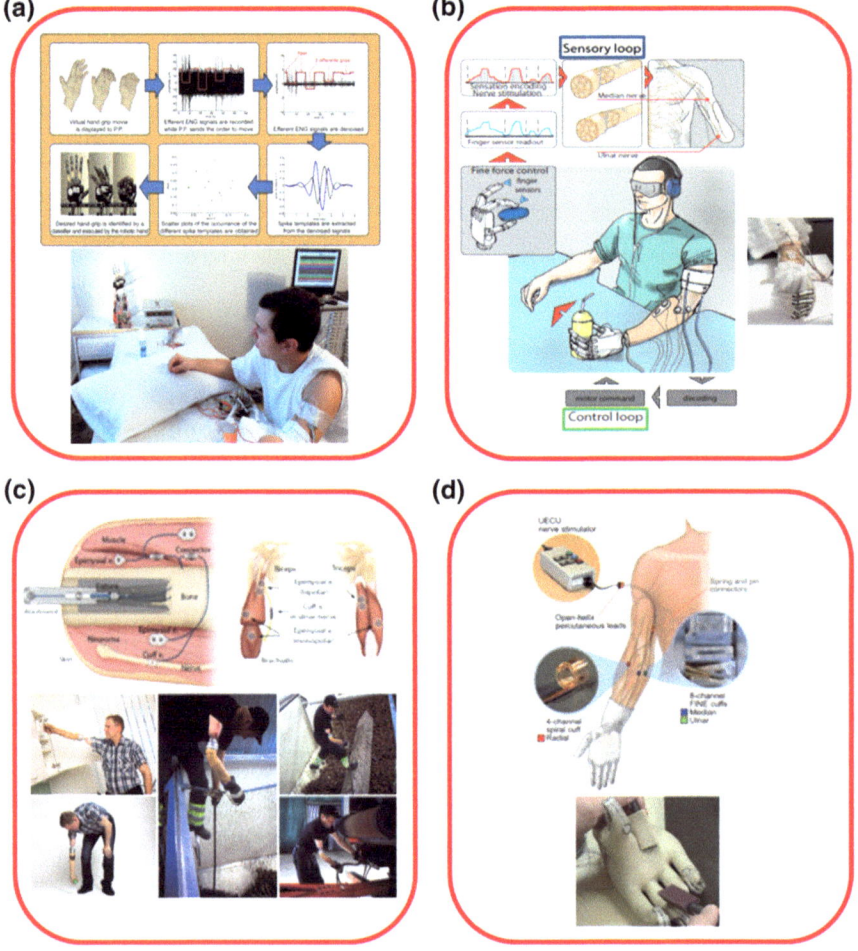

**Fig. 5** Experimental studies on the restoration of the natural tactile sensory feedback stimulating the nerve in upper-limb amputees: **a** Rossini et al. (2010), **b**) Raspopovic et al. (2015), **c** Ortiz-Catalan et al. (2014), **d** Tan et al. (2014)

elicit sensations (parameters in Table 4). A mapping phase of the 32 electrodes allows identifying the elicited sensations on the afferent fibers (Rossini et al. 2010) (Fig. 6). Different contacts permitted to elicit different sensations. Moreover, the improvement of the phantom limb pain symptoms has been observed by means of McGill Pain Questionnaire (sfMcGill), Present Pain Intensity (PPI) and Visual Analogue Scale (VAS).

The work in (Raspopovic et al. 2015) reported the results of a surgical implant of four transversal intrafascicular multichannel electrodes (TIMEs, Boretius et al. 2010), two in median nerve and two in ulnar nerve, for the bidirectional control of the hand prosthesis. The control has been achieved by means of a myoelectric control of the prosthesis on the efferent pathway and a sensory loop that, reading the hand sensor readouts, was able to elicit tactile sensations on the afferent pathway.

In the first phase, a mapping of all contact sites allowed identifying all the possible elicitable sensations and the related territories. In the second phase, the closed-loop control around the user has been experimented by decoding the user intention. The Prensilia IH2 Azzurra hand provided with tension sensors on the tendons has been used during the experiments.

Figure 2b reports the hand areas elicited from the electrical stimulation of median and ulnar nerves. The amputee has been able to voluntarily control three levels of pressure exerted on the object with index and little fingers with a success

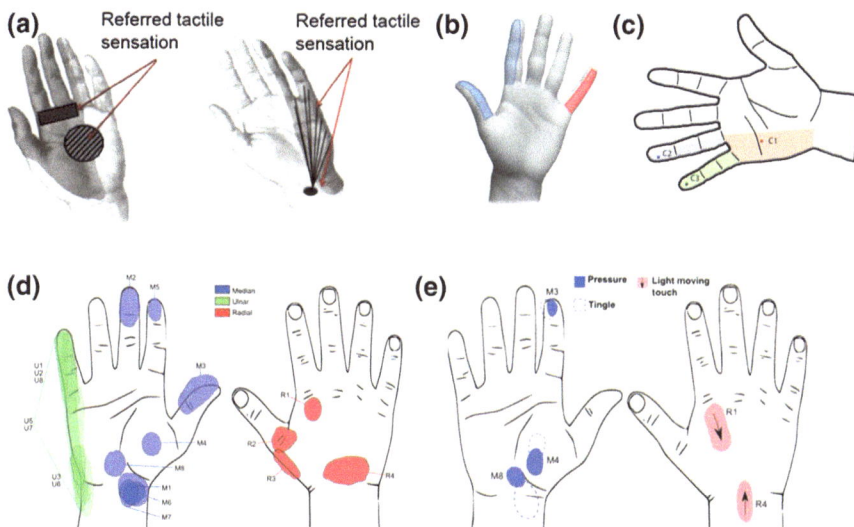

**Fig. 6** **a** Perceived localization of sensation after median nerve tf-LIFE4 stimulation on left and after ulnar nerve tf-LIFE4 stimulation on right (Rossini et al. 2010); **b** Elicited hand areas in Raspopovic et al. (2015); **c** Tactile perception in Ortiz-Catalan et al. (2014). The *dark points* represent the electrode-specific projected field; **d** Sensation locations (Tan et al. 2014). The *letter* represents the nerve and the *number* represents the stimulus channel; **e** Pressure tactile perception on varying of impulse duration (Tan et al. 2014)

rate >90%. Moreover, the subject has been able to identify three object shapes with mean accuracy of 88% and three different consistency levels with performance of 78.7% (Raspopovic et al. 2015).

The study in Ortiz-Catalan et al. (2014) demonstrates the improvement of hand controllability using implanted EMG sensors and the feasibility of eliciting tactile sensations in chronic implants. One cuff electrode has been permanently implanted in the ulnar nerve of one patient with transhumeral amputation treated with an osteointegration procedure.

The controllability of the prosthesis has been improved and the time of usage of the robotic hand increased of 6 h per day respect to the use of superficial EMG sensors. The results achieved with the prosthetic hand are comparable with those obtained with the healthy hand. Eight different movements (hand opening/closing, wrist pronation/supination, wrist and elbow flexion/extension) have been performed with an accuracy of 94.3% ($\sigma = 1.6\%$).

The evolution of the phantom limb pain and the use of the prosthesis have been evaluated through the use of sfMcGill measuring a decrease of 40% in the phantom limb pain between two months before and 10–16 months after the implant.

Finally, the study in Tan et al. (2014) shows the feasibility of eliciting tactile sensations in a stable manner up to 24 months and demonstrates that the perceived sensations and the perceived areas can be modulated changing pulses parameters (Table 4).

Two subjects have been implanted with FINE (flat interface nerve electrodes) (Tyler and Durand 2003) cuffs or with CWRU (Case Western Reserve University) spiral electrode (1988). The first patient has been implanted with two FINE cuffs with eight contacts in median and ulnar nerves and one CWRU spiral electrode with four contacts in radial nerve, with totally 20 active sites. The second subject has been provided with totally 16 active sites: two FINE cuffs with eight contacts implanted in median and radial nerves.

The control of a prosthetic hand taking advantages from the sensory feedback has been investigated in (Schiefer et al. 2015). The EMG signals have been used to control the SensorHand Speed provided with FlexiForce sensors on thumb and index tips.

The impact of the sensory feedback on the prosthesis use has been assessed by means of three different functional tests. The first test aimed at evaluating the subject's ability to distinguish the position of a wooden block during index/thumb and middle/thumb pinch grip. The second test, a modified version of the box and blocks test, has been performed to verify if the amputee was able to locate and move a block. Finally, the amputee ability to perform ADLs (Schiefer et al. 2015) has been evaluated by means of the southampton hand assessment procedure (SHAP) applied with and without sensory feedback. Despite the SHAP score has been improved with sensory feedback, mainly during Power, Tip, and Lateral grasps, the assessment highlighted for both subjects a major focus on visual feedback than tactile feedback.

The analyzed studies provide an evidence of the possibility of using afferent neural stimulation to elicit sensory feedback and establish a bidirectional control with the user. Notwithstanding the reviewed studies have to be acknowledged for the contribution they have brought in the prosthetic field, a number of challenges still need to be faced by the future research studies. For instance, current prostheses do not fully exploit hardware potentiality they embed to perform complex grasping and fine manipulation tasks. More advanced solutions for sensorization and control of upper-limb prostheses should be developed. Moreover, performance of the prosthetic device is still too dependent on the interfacing system and its limitations. Finally, biostability and reliability of neural interfaces need to be further investigated for clinical translation of currently achieved preliminary scientific results and, possibly, less invasive solutions for the restoration of close-to-natural sensations need to be explored. Most of these challenges are faced in the currently active project PPR2—control of upper-limb prosthesis by invasive neural interfaces—involving the authors of this chapter and supported by Italian Institute for Insurance against Accidents at Work (INAIL).

## 6 Conclusions

This chapter has provided an overview of the main achievements and current trends in the field of upper-limb prostheses.

Requirements coming from the analysis of users' needs have been presented. They mainly regard the hand dexterity and sensory feedback and can be satisfied by working on the prosthesis control and the recovery of the bidirectional communication with the PNS through a closed-loop interface.

Hence, special attention has been devoted to the neural interfaces for prosthesis control and afferent stimulation and to the neural implants for the restoration of sensory feedback. This chapter also analyzes the most eminent human experiments on bidirectional hand prostheses that close the user in the control loop by means of neural interfaces.

Notwithstanding the encouraging results, current solutions are still far from being translated into clinical practice for a number of open issues regarding invasiveness, reliability, robustness, long-term stability. The deployment of the discussed solutions in the clinical practice mainly depends on the stability and reliability of the developed hardware and software solutions over long periods of time. This chapter reveals the necessity to improve hand control and the interfaces between the prosthesis and the human body in order to restore in a reliable and robust way the amputee sensation during interaction with environment and the way of exchanging sensory information between the environment and the user. Future challenges will be focused on (a) the increase of dexterity of currently available prosthetic hands, working on control and sensory system, (b) the improvement of tactile selectivity and discrimination capabilities by combining multimodal sensory system intraneural with peripheral interfaces, and (c) the development of biostable and biocompatible neural interfaces able to permanently restore the bidirectional communication with the PNS.

**Acknowledgements** This work was supported by the Italian Institute for Labour Accidents (INAIL) with PPR 2 project (CUP: E58C13000990001) and by the European Project H2020/AIDE (CUP J42I15000030006).

# References

Antfolk, C., D'Alonzo, M., Rosen, B., Lundborg, G., Sebelius, F., & Cipriani, C. (2013). Sensory feedback in upper limb prosthetics. *Expert Review of Medical Devices, 10,* 45–54. doi:10.1586/erd.12.68.

Badia, J., Boretius, T., Andreu, D., Azevedo-Coste, C., Stieglitz, T., Navarro, X. (2011). Comparative analysis of transverse intrafascicular multichannel, longitudinal intrafascicular and multipolar cuff electrodes for the selective stimulation of nerve fascicles. *Journal of Neural Engineering, 8,* 13 pp.

Belter, J. T., Segil, J. L., Dollar, A. M., & Weir, R. F. (2013). Mechanical design and performance specifications of anthropomorphic prosthetic hands: A review. *Journal of Rehabilitation Research and Development, 50,* 599–618.

Benvenuto, A., Raspopovic, S., Hoffmann, K. P., Carpaneto, G. C., Di Pino, G., Guglielmelli, E. (2010). Intrafascicular thin-film multichannel electrodes for sensory feedback: Evidences on a human amputee. In *32nd Annual International Conference of the IEEE EMBS*, pp. 1800–1803.

Biddiss, E., Beaton, D., & Chau, T. (2007). Consumer design priorities for upper limb prosthetics. *Disability and Rehabilitation: Assistive Technology, 2,* 346–357. doi:10.1080/17483100701714733.

Biddiss, E. A., & Chau, T. T. (2007). Upper limb prosthesis use and abandonment: A survey of the last 25 years. *Prosthetics and Orthotics International, 31,* 236–257. doi:10.1080/03093640600994581.

Boretius, T., Badia, J., Pascual-Font, A., Schuettler, M., Navarro, X., Yoshida, K., et al. (2010). A transverse intrafascicular multichannel electrode (TIME) to interface with the peripheral nerve. *Biosensors & Bioelectronics, 26,* 62–69.

Bossi, S., Menciassi, A., Koch, K. P., Hoffmann, K. P., Yoshida, K., Dario, P., et al. (2007). Shape memory alloy microactuation oft f-LIFEs: Preliminary results. *IEEE Transactions on Biomedical Engineering, 54*(6), 1115–1120.

Bouffard, J., Vincent, C., Boulianne, E., Lajoie, S., & Mercier, C. (2012). Interactions between the phantom limb sensations, prosthesis use, and rehabilitation as seen by amputees and health professionals. *Journal of Prosthetics and Orthotics., 24,* 25–33. doi:10.1097/JPO.0b013e318240d171.

Childress, D. S. (1980). Closed-loop control in prosthetic systems: Historical perspective. *Annals on Biomedical Engineering, 8,* 293–303.

Ciancio, A. L., Cordella, F., Barone, R., Romeo, R. A., Dellacasa Bellingegni, A., Sacchetti, R. et al. (2016). Control of prosthetic hands via the peripheral nervous system. *Frontiers in Neuroscience, 10,* 116.

Clement, R. G. E., Bugler, K. E., & Oliver, C. W. (2011). Bionic prosthetic hands: A review of present technology and future aspirations. *Surgeon, 9,* 336–340.

Clippinger, F. W., Avery, R., & Titus, B. R. (1974). A sensory feedback system for an upper-limb amputation prosthesis. *Bulletin of Prosthetics Research*, 247–258.

Cloutier, A., Yang, J. (2013). Design, control, and sensory feedback of externally powered hand prostheses: A literature review. *Critical Reviews in Biomedical Engineering, 41,* 161–81. doi:10.1615/CritRevBiomedEng.2013007887.

Cloutier, A., & Yang, J. (2013). Design, control, and sensory feedback of externally powered hand prostheses: A literature review. *Critical Reviews in Biomedical Engineering, 41*, 161–81. doi:10.1615/CritRevBiomedEng.2013007887.

Cogan, S. F. (2008). Neural stimulation and recording electrodes. *Annual Review of Biomedical Engineering*, 275–309.

Cordella, F., Ciancio, A. L., Sacchetti, R., Davalli, A., Cutti, A. G., Guglielmelli, E., et al. (2016). Literature review on needs of upper limb prosthesis users. *Frontiers in Neuroscience.* doi:10.3389/fnins.2016.00209.

Dhillon, G. S., & Horch, K. W. (2005). Direct neural sensory feedback and control of a prosthetic arm. *IEEE Transactions on Neural Systems and Rehabilitation Engineering, 13*, 468–472. doi:10.1109/TNSRE.2005.856072.

Dohnalek, P., Gajdos, P., & Peterek, T. (2013). Human activity recognition on raw sensors data via sparse approximation. In *Proceedings of the 36th International Conference on Telecommunications and Signal Processing*, pp. 700–703. doi:10.1109/TSP.2013.6614027.

Farina, D., & Aszmann, O. (2014). Bionic limbs: Clinical reality and academic promises. *Science Translational Medicine, 8*; 6(257):257, 12, 1.O.C.

Fougner, A., Stavdahl, O., Kyberd, P. J., Losier, Y. G., & Parker, P. (2012). Control of upper limb prostheses: terminology and proportional myoelectric control: A review. *IEEE Transactions on Neural Systems and Rehabilitation Engineering, 20*, 663–677.

Fumero, R., Costantino, M. L. (2001). Organi artificiali. In *Storia della Bioingegneria* (vol. 6, pp. 341–365). Patron.

Gonzales, D. S., & Castellini, C. (2013). A realistic implementation of ultrasound imaging as a human-machine interface for upper limb amputees. *Frontiers in Neurorobotics, 7*, 1–11. doi:10.3389/fnbot.2013.00017.

Hoffmann, K. P., & Kock, K. P. (2005). Final report on design consideration of tLIFE2. Technology Report IBMT.

Hoffmann, K. P., & Micera, S. (2011). Neuroprosthetics. In R. Kramme, K. P. Hoffmann, & B. Pozos (Eds.), *Handbook medical technology* (1st ed., pp. 785–800). Heidelberg New York: Springer.

Jang, C. H., Yang, H. S., Yang, H. E., Lee, S. Y., Kwon, J. W., Yun, B. D. et al. (2011). A survey on activities of daily living and occupations of upper extremity amputees. *Annals of Rehabilitation Medicine*, 35. doi:10.5535/arm.2011.35.6.907.

Jensen, W., Micera, S., Navarro, X., Stieglitz, T., Guiraud, D., Divoux, J. L. et al. (2010). Development of an implantable transverse intrafascicular multichannel electrode (TIME) system for relieving phantom limb pain. In *32nd Annual International Conference of the IEEE EMBS*, pp. 6214–6217.

Jiang, N., & Farina, D. (2014). Myoelectric control of upper limb prosthesis: Current status, challenges and recent advances. Front. Neuroeng. Conference.

Kyberd, P. J., & Hill, W. (2007). Survey of upper limb prosthesis users in Sweden and the United Kingdom. *Journal of Prosthetics and Orthotics, 19*, 55–66. doi:10.1177/0309364611409099.

LeBlanc, M. (2008). Give hope—Give a hand. *The LN-4 prosthetic hand.* Accessed October 1, 2013, Available:http://www.stanford.edu/class/engr110/2011/LeBlanc-03a.pdf.

Li, N., Yang, D., Jiang, L., Liu, H., & Cai, H. (2012). Combined use of FSR sensor array and SVM classifier for finger motion recognition based on pressure distribution map. *Journal of Bionic Engineering, 9*, 39–47. doi:10.1016/S1672-6529(11)60095-4.

Lucchetti, M., Cutti, A. G., Verni, G., Sacchetti, R., & Rossi, N. (2015). Impact of Michelangelo prosthetic hand: Findings from a crossover longitudinal study. *Journal of Rehabilitation Research and Development, 52*(5), 605–618. doi:10.1682/JRRD.2014.11.0283.

Micera, S., Navarro, X., Carpaneto, J., Citi, L., Tonet, O., Rossini, P. M., et al. (2008). On the use of longitudinal intrafascicular peripheral interfaces of the control of cybernetic hand prostheses in amputees. *IEEE Transactions on Neural Systems and Rehabilitation Engineering, 16*(5), 453–472.

Naples, G. G., Mortimer, J. T., Scheiner, A., & Sweeney, J. D. (1988). A spiral nerve cuff electrode for peripheral nerve stimulation. *IEEE Transactions on Biomedical Engineering, 35*, 905–916.

Navarro, X., Krueger, T. B., Lago, N., Micera, S., Stieglitz, T., & Dario, P. (2005). A critical review of interfaces with the peripheral nervous system for the control of neuroprostheses and hybrid bionic systems. *Journal of the Peripheral Nervous System, 10*, 229–258.

Navarro, X., Lago, N., Vivó, M., Yoshida, K., Koch, K. P., Poppendieck, W., et al. (2007). Neurobiological evaluation of thin-film longitudinal intrafascicular electrodes as a peripheral nerve interface. In *Proceedings of the 2007 IEEE 10th International Conference on Rehabilitation Robotics*, pp. 643–649.

Ortiz-Catalan, M., Hakansson, B., & Branemark, R. (2014). An osseointegrated human-machine gateway for long-term sensory feedback and motor control of artificial limbs. *Science Translational Medicine*. doi:10.1126/scitranslmed.3008933.

Østlie, K., Lesjø, I. M., Franklin, R. J., Garfelt, B., Skjeldal, O. H., & Magnus, P. (2012). Prosthesis use in adult acquired major upper-limb amputees: Patterns of wear, prosthetic skills and the actual use of prostheses in activities of daily life. *Disability and Rehabilitation: Assistive Technology, 7*, 479–493. doi:10.3109/17483107.2011.653296.

Pasquina, P. F., Evangelista, M., Carvalho, A. J., Lockhart, J., Griffin, S., Nanos, G., et al. (2015). First-in-man demonstration of a fully implanted myoelectric sensors system to control an advanced electromechanical prosthetic hand. *Journal of Neuroscience Methods, 244*, 85–93. doi:10.1016/j.jneumeth.2014.07.016.

Peerdeman, B., Boere, D., Witteveen, H. J. B., Hermens, H. J., Stramigioli, S., Rietman, J. S., et al. (2011). Myoelectric forearm prostheses: State of the art from a user-centered perspective. *Journal of Rehabilitation Research and Development, 48*, 719–737. doi:10.1682/JRRD.2010.08.0161.

Popov, B. (1965). The bio-electrically controlled prosthesis. *The Journal of Bone and Joint Surgery (British), 47B*, 421–424.

Poppendieck, W., Muceli, S., Dideriksen, J., Rocon, E., Pons, J. L., Farina, D. et al. (2015). A new generation of double-sided intramuscular electrodes for multi-channel recording and stimulation. 37th Annual International Conference of the IEEE EMBS, pp. 7135–7138.

Pylatiuk, C., Schultz, S., & Doderlein, L. (2007). Results on internet survey of myoelectricprosthetic hand users. *Prosthetics and Orthotics International, 31*(4), 362–370. doi:10.1080/03093640601061265.

Raspopovic, S., Capogrosso, M., Petrini, F., Bonizzato, M., Rigosa, J., Di Pino, G. et al. (2015). Restoring natural sensory feedback in real-time bidirectional hand prostheses Science. *Translational Medicine, 6*(222).

Rossini, P. M., Micera, S., Benvenuto, A., Carpaneto, J., Cavallo, G., Citi, L., et al. (2010). Double nerve intraneural interface implant on a human amputee for robotic hand control. *Clinical Neurophysiology, 121*, 777–783.

Schiefer, M., Tan, D., Sidek, S. M., & Tyler, D. J. (2015). Sensory feedback by peripheral nerve stimulation improves task performance in individuals with upper limb loss using a myoelectric prosthesis. *Journal of Neural Engineering, 13*(1), 016001.

Schofield, J. S., Evans, K. R., Carey, J. P., Hebert, J. S. (2014). Applications of sensory feedback in motorized upper extremity prosthesis: A review. *Expert Review of Medical Devices*, 1–13.

Tan, D. W., Schiefer, M. A., Keith, M. W., Anderson, J. R., Tyler, J., & Tyler, D. J. (2014). A neural interface provides long-term stable natural touch perception. *Science Translational Medicine, 6*, 257ra138. doi:10.1126/scitranslmed.3008669.

Tyler, D. J., & Durand, D. M. (2002). Functionally selective peripheral nerve stimulation with a flat interface nerve electrode. *IEEE Transactions on Neural Systems and Rehabilitation Engineering, 10*(4), 294–303.

Tyler, D. J., & Durand, D. M. (2003). Chronic response of the rat sciatic nerve to the flat interface nerve electrode. *Annals of Biomedical Engineering, 31,* 633–642. doi:10.1114/1.1569263.

Van Lunteren, A., Van Lunteren-Gerritsen, G. N. M., Stassen, N. C., & Zuithoff, M. J. (1983). A field evaluation of arm prostheses for unilateral amputees. *Prosthetics and Orthotics International, 7,* 51–141. doi:10.3109/03093648309166586.

Wininger, M., Kim, N. H., Craelius, W. (2008). Pressure signature of fore-arm as predictor of grip force. *Journal of rehabilitation research and Development, 45,* 883–892. doi:10.1682/JRRD.2007.11.0187.

Wright, T. W., Hagen, A. D., & Wood, M. B. (1995). Prosthetic usage in major upper extremity amputations. *Journal of Hand Surgery., 20A,* 22–619. doi:10.1016/S0363-5023(05)80278-3.

# Part II
# The Hand, Human Identity and Creativity

# The Human Hand as a Microcosm. A Philosophical Overview on the Hand and Its Role in the Processes of Perception, Action, and Cognition

Maria Teresa Russo

> Here lies the body of Lois Spears, …Children with clear eyes and sound limbs—(I was born blind). I was the happiest of women As wife, mother and housekeeper, Caring for my loved ones, And making my home A place of order and bounteous hospitality: For I went about the rooms, And about the garden With an instinct as sure as sight, As though there were eyes in my finger tips
> 
> E. Lee Masters, *Spoon River Anthology* (1916), *Lois Spears*

**Abstract** The phenomenological analysis of the hand is a fertile ground to reflect on the dialogue between philosophy and the biomedical sciences. In fact, this analysis highlights the correspondence which exists in human beings between body morphology and symbolic intelligence. In ancient philosophy, there is a confrontation between the mechanicistic thesis by Anaxagoras and the finalistic thesis by Aristotle. Starting from the hand, the argument extends to the role that touch has in relation to the other senses within perception and self-consciousness. The data of developmental psychology and neurosciences confirm that self-consciousness is growing gradually by learning to distinguish between what is own and what is extraneous, thanks to tactile sensations. In contemporary philosophy, we have the confrontation of two views: one is focused on haptic experience which attributes to touch a central role in the perceptive process and in the interpersonal relation, according to a line that goes from Aristotle to Kant and further on to Husserl, Derrida, and Lévinas. The other position considers sight as the more capable faculty to recognize and to imagine: this is the position of Merleau-Ponty who identifies the tangible with the visible. The question emerging from this confrontation—"is it the look that touches or the hand that sees?"—may find a practical reply within cognitive processes of persons suffering from congenital blindness.

**Keywords** Embodiment · Gesture · Philosophical Anthropology · Perception · Phenomenology · Touch

M.T. Russo (✉)
Roma Tre University, Rome, Italy
e-mail: mariateresa.russo@uniroma3.it

© Springer International Publishing AG 2017
M. Bertolaso and N. Di Stefano (eds.), *The Hand*, Studies in Applied Philosophy, Epistemology and Rational Ethics 38, DOI 10.1007/978-3-319-66881-9_6

## 1 Morphology of the Human Body, Symbolic Intelligence, and Hand

The phenomenological analysis of the hand is a fertile ground to reflect on the dialogue between philosophy and the biomedical sciences. It is exactly in this analysis, in fact, that the correspondence between human body morphology and symbolic intelligence becomes evident. This correspondence appears in the systemic character of the human body, where all the parts are functionally interconnected and interdependent, and in the so-called bodily "intentionality" which allows the human being to carry out all the different actions, even those not directed at survival. Indeed, the more perfect the living being is, the more complex its bodily structure is too. In the case of a human being, the maximum perfection of the body is evident not simply in the complexity of its parts, but especially in their reciprocal correlation and interaction. In human morphology, the systemic feature, trait present in every living being, is at its maximum, showing a special connection among the different parts and a direct correspondence between cognitive capacities and the ability to exploit them. Hand and face, for instance, from an expressive point of view constitute a system, and the same can be said for arm and hand.

In ancient philosophy, there is a debate between the mechanicistic thesis by Anaxagoras, who argued that the human being is the most intelligent of the animals because of his hands,[1] and the finalistic thesis by Aristotle, who stated, instead, that intelligence is the reason why we have hands, and not vice versa. The thesis argued by Anaxagoras is still present in the different anti-finalisms that explain structure through function. According to Aristotle, and then to other later thinkers, the function, in this case, the multitasking ability of the hand, depends instead on the human rationality.

If the hand is a tool, we need instrumental abilities to use it: the hand is the way it is because there is an intelligence that knows how to use it.[2]

The hand has an operative opening, which means it can be used in many different ways not necessarily and exclusively linked to survival: grasp, feel, manipulate, but also point, clap, play. In addition to this, fingers, which function as an integrate sensory-motor unit, allow us to recognize objects by understanding shapes. Manipulation has, therefore, a cognitive function which requires a sort of mental distancing, a "disinterested interest" that characterizes knowledge as an opening to the world, not simply in an instrumental or technical way but also in a reflective one. Aristotle defines manipulation as the "instrument of instruments"[3] and highlights its polytechnic and not-specialized nature. Upright standing and bipedalism made it possible to free front limbs and therefore it allowed functions of manipulation. The anatomic difference between the front and the rear limbs

---

[1]Cfr. H. Diels—W. Kranz, Die Fragmenten der Vorsokratiker, Weidmann, Dublin-Zürich 1972$^{16}$, unveränderte Nachdruck ver 6. Auflage, A 102.

[2]Aristotle, *De partibus animalium* IV 10, 687 a 9–11.

[3]Aristotle, *De Anima* III 8, 432 a 1–3.

extremities is an important milestone in the hominization process. There are two structural features of the human hand that makes the prehensile function possible: the convergence–divergence and the thumb opposition. The former allows both finger flexion and adduction, and their fan-shaped opening. The latter enables the thumb to place itself at a 90-degree angle with respect to the other fingers.[4]

As it has been shown, a fundamental difference between human and other primates' hand lies in the existence of two apparently opposite possibilities to grip: the strength, which allows to wring and squeeze or press, and the precision of the pincer grip through which, for instance, the human being can thread a needle.[5] This makes the human hand a tool that has the ability to "endlessly use finite resources".[6] It is interesting to note that it is not the object, or at least not only the object, that determines the type of grip needed but the goal of the action. The newborn baby has only the strength to grip, and we need to wait for the baby to be one-year old to see the precision grip too. Sennett calls "minimal force"[7] the precision ability, which both cooks and surgeons have and which is closely related to the ability to let the grip go,[8] essential, for example, to a pianist. In these movements, the hand is part of a motor and perceptual system which includes the muscles and the joints of the arm and of the forearm, but also the skin.

Upright standing does not simply offer the possibility to use hands, but it also provides that psychological experience that Lewin defines as "space of action" or "hodological space".[9]

Differently from the space of the Cartesian science, which is neutral and homogeneous, the hodological space is lived, in relation to its own body and to the possibility to move in it and to inhabit it. In this space, hands become organs of perception, tools of "gnostic touch" that together with vision and hearing enable recognition and distinction. Human life develops in fact in a spatial dimension, where distance and physical contact play a fundamental role. In this case, everything that can be reached by our own body becomes "handy"; at the same time, the affective perception of the space—comfortable or uncomfortable—is connected to an equilibrium between proximity and distance. Think about the problems due to crowds, which makes contact inevitable and uncomfortable.

Tactile space is a dimension where *texture*, that is the quality of the materials that can be perceived through touch, plays an important role too. Smooth, rough, velvety, prickly, hard, soft are characteristics that have a direct influence not only on esthetic perception but also on the experience of feeling good in a certain

---

[4]Révèsz (1950), Gibson(1966).
[5]Napier (1956), Connally and Elliott (1972), Tallis (2003).
[6]Tallis (2003)
[7]Sennett (2008).
[8]Sennett, pp. 148–149.
[9]Lewin (1934).

place.[10] Japanese culture is among the most sensitive to this dimension, thanks to a special attention to the choice of materials for buildings, floors, and walls.

Texture is also involved in taste perception: this clearly shows the unity of the sensory and of the perceptual systems, which are closely linked to the activation of emotive reactions of attraction and repulsion.

The hand has also a symbolic function exactly because of the expressive nature of the human body, which manifests itself through gestures. Culture attributes even communication qualities to each finger: the thumb pointing down convicts; the index admonishes and assigns; the middle finger is rude and indecent; index and middle fingers stand for victory. Oaths and greetings involve the use of hands in various ways.[11] The symbolic nature of hands and touch is recognized by common expressions in different languages: "have someone in hand," "handle a situation," "give someone a hand."

## 2 The *Intelligent Hand*: Perceptual and Cognitive Aspects

The density of skin receptors on the hand varies from one area to another. The back of the hand is less sensitive than the palm, and the phalanxes, as well as each finger, have different sensitivities. For this reason, the manual perception of the dimension and shape of an object is different if the subject explores it with the entire hand or only with the extremities of the fingers, or with one or more fingers.

Davidson, comparing the behavior of sighted people and blind, distinguishes five techniques of manual exploration[12]: *grip*, when the grasp happens with three or four fingers on the external part of the object; *pinch*, if the thumb and another finger touch the vertical part of an object; *topsweep*, when a single rigid finger rubs the upper part of an object multiple times; *span*, if you measure the object with the entire hand; *trace*, if you follow the internal side of the object with a stretched finger. We need to add another property: the coordination between the two hands and, at the same time, their independence from each other.[13] Bimanual perception has nothing in common with the binocular one. While eyes cannot look in two different directions, hands are characterized by a functional asymmetry that increases the motor independence: hands can execute identical or different, synchronous or alternate movements.

Finally, we need to underline the "metric function" of the hand, which derives exactly from the morphology of the palm and of the fingers that are natural meters.[14] In fact, the first units of measure have been parts of the body: feet, inches, etc.

---

[10]Cfr. Hall (1966).
[11]Le Breton (2006), Paterson (2007)
[12]Davidson (1972).
[13]p. 159.
[14]Révèsz (1950)

What is the relationship between manual practice and cognitive development? Culture has often valued the development of sight and of intellectual capacities more than manual exercise. Yet there are theories that have instead promoted manual skills. In the history of pedagogy, the method devised by Maria Montessori, based on the thesis that "the hand is the organ of intelligence," puts the promotion of manipulative skills at the center, believing that if we train kids in working with their hands, we promote their cognitive development and a balanced personality.[15] There are two more examples in which the fundamental cognitive function of manual skills is evident: calculating with fingers and manual writing. As the most recent neuroscientific studies show, preventing kids from counting with their fingers does not promote the development of their ability to count because there is a direct link between the execution of calculations and the activation of the somatosensory cortex, the area of the brain for hand and tactile sensations' representations.

Accordingly, the better the representation of the fingers one has, the better the mathematical results.[16] On the contrary, the "digital agnosia," that is the bad representation of one's own fingers, is a good predictor of dyscalculia.[17]

Another field in which teaching manual skills has a fundamental relevance is writing, which is the result of a complex cognitive process and the synthesis of intellectual, perceptive, and motor abilities. Recent studies, mainly linked to the development of research projects on *embodied cognition*, focused on the writing process and on its importance for cognitive development, which had previously been analyzed from a theoretical–methodological perspective and which had anyway been neglected in favor of literature.[18] Today, the interest is greater also because of the development of digital writing, which led many scholars to ask questions about the transformations that the decline of handwriting practice can cause. The comparison between the features of the two practices—handwriting and digital writing—is interesting.[19] The first entails a slower process, characterized by a unimanual activity and by the fact that the attention is focused on the use of the pen, to produce a graphic shape as close as possible to the standard shape of each letter (grapho-motor component). It is, therefore, necessary to integrate visual, haptic-kinesthetic, and tactile data to achieve the synthesis of the perceptive element (learning the shape of every letter) and of the grapho-motor element (learning the trajectory from which the shape of each letter derives). Digital writing is instead a

---

[15]"He does it with his hands, by experience, first in play and then through work. The hands are the instruments of man's intelligence". Montessori (1949, p. 25); "Man builds himself through working, working with his hands, but using his hands as the instruments of his ego, the organ of his individual mind and will, which shapes its own existence face to face with its environment". Montessori (2006, p. 195).

[16]Butterworth (2005).

[17]Berteletti and Booth (2015), Gracia-Bafalluy and Noël (2008), Newman and Soylu (2013).

[18]See the publications and the activities of the *National Handwriting Association*: http://www.nha-handwriting.org.uk/ and the advertising campaign of *Bic Pen* "Fight for your write": https://www.bicfightforyourwrite.com/.

[19]Mangen and Velay (2010).

faster process, characterized by a bimanual activity and by the fact that the attention is focused on the keyboard or on the screen; this process is divided into two separate phases from the spatiotemporal point of view: the motor phase and the visual phase. The grapho-motor component is missing since the fonts are pre-formed and therefore there is no difference between the graphic shapes of a beginner and those of an expert writer. The result is a fundamental change not only in the hands' role and, more in general, in the tactile perception but also in the linguistic and mnestic skills.[20]

Also in the field of architecture, there have been concerns about the predominance of computerized drawing, the so-called computer-aided design (CAD), which has almost completely replaced manual drawing. By removing the hand, the computer creates a distance between the designer and the object, resulting in a reduced multisensorial imagination synchronized only with the visual dimension, surely more accurate but uncoupled from concrete reality.[21] Sennett points out three risks derived from the abuse of CAD and the lacking of hand drawing: The disconnect between simulation and reality; the loss of a certain kind of relational understanding; finally, CAD's precision brings out a problem of over determination.[22]

## 3 Is the Hand Seeing or the Eye Touching? The Investigation of Philosophy

If the hand enables us to know and to dwell in the world, the fragmentation of the object caused by each manual fixation makes tactile perception sequential compared to the visual one, which instead catches at the same time the position in space. In the philosophical tradition, we therefore witness a debate between two conceptions: the prevailing one is *oculocentric*, which separates the "theoretical" senses (sight and hearing) from the "not theoretical" ones (taste, touch, sense of smell) to assign the primacy to the first ones, and to sight in particular.[23] Hence, the importance attributed to light, which becomes the tool par excellence through which sight can reach the object. The other one, instead, by promoting the integral bodily experience, restores the importance of the hand and of touch more in general. This recurrent dichotomy between hand and eye constitutes, from Descartes onwards, one of the fundamental pillars of the reflection on sensitivity and, all things considered, on the status of thought itself.

---

[20]Kress (2003), Berninger et al. (2006).
[21]Pallasmaa (2009).
[22]Sennett (2008), p. 59.
[23]"Eyes are more precise witnesses than ears" (fr. 101a) says one of Heraclitus' fragments (sec. V a. C.). Plato, in the *Teeteto*, defines sight, sensation and knowledge as "the same thing". Classen (2012).

According to Aristotle, touch is the only sense necessary for the existence of living beings, because it is linked to nutrition[24]: the other senses are not directed at assuring the being of the animal or of the human being, but only at its *well*-being.[25] Nevertheless, he stresses its problematic nature since, differently from the other four senses that have a unique own sensitivity, touch is about a multiplicity of objects, even opposite to each other: cold and hot, dry and humid, hard and soft. In addition to this, touch does not have a medium, since the tactile sensation is given via a direct and immediate contact between the object and the subject. The philosopher comes to the conclusion that the medium is the flesh itself so that the organ of touch is internal because it cannot identify itself with the medium (that is with the flesh).

Touch has also two more peculiarities: the medium can be perceived together with the object, as it happens when one gets hit through a shield; the shield, in fact, does not hit the person after being beaten, but both get hit simultaneously. Finally, the four qualities of touch get perceived in relation to the quality of the subject: to perceive something hot, for instance, we need to be less hot than the perceived object; otherwise, we would not be able to perceive it.[26]

Aristotle seems to contradict himself: on one hand he believes touch is the most acute sense that humans have, followed by taste, while for the other senses, humans are behind many other animals[27]; on the other hand, he states that sight has the maximum acuteness.[28] Indeed, in the first case, the philosopher is establishing a comparison between human touch and that of other animals, while in the other texts he defines which sense, among those available to humans, can be said to be the most acute one. For this reason, the importance attributed to touch it is not an epistemological one: touch is the only sense necessary and sufficient to define an animal. Even if it is the most universal one, its function is inferior to that of hearing —which is the most speculative sense—and especially to that of sight, which allows to better know all the other common sensibles (movement, rest, number, figure, size) and, for this reason, is the most useful sense for daily life needs.[29]

The supremacy of sight is also due to its theoretical and detached character. As it is argued in Metaphysics, sight is "perfect action," which opens us to knowledge more than the other senses.[30]

---

[24]Aristotle, *De Anima* II, 413b–414 b 5–10.

[25]Ivi.

[26]Aristotle, *De anima* II 11, 423 b 17–20.

[27]Aristotle, *Historia animalium*. I 15, 494 b 16–18.

[28]Aristotle, *De partibus animalium* 656 a 35–b 6; *De* sensu 1, 437 a.

[29]Talking about virtues, Aristotle attributes intemperance to tactile sensations, since the pleasures provided by sight, hearing and smell do not admit excesses. Aristotle, *Etica Eudemia* III, 1 1230b–1231a.

[30]"All men by nature desire to know. An indication of this is the delight we take in our senses; for even apart from their usefulness they are loved for themselves; and above all others the sense of sight. For not only with a view to action, but even when we are not going to do anything, we prefer sight to almost everything else. The reason is that this, most of all the senses, makes us know and brings to light many differences between things". Aristotle (1960).

Even Descartes, in the opening of his discourse in *Optics*, describes sight as "the noblest and most comprehensive of the senses," considering touch and taste as the most coarse ones.[31]

On the other hand, in his rationalist perspective, data provided by the senses are variables and do not help us understand the real essence of things, rather they only inform us on what is useful or dangerous. Truth needs to be found via an investigation of thought that goes beyond the façade of sensory data since there are no sensations independent of *res cogitans*.

The construction of optical instruments will stimulate during the sixteenth century the development of optical theories on the function of sight. The question that William Molyneux, astronomer, physicist, and member of the *Trinity College* in Dublin, asks to his friend in a letter dated March 2, 1693, is meaningful: a person born blind who, thanks to touch, has learnt to distinguish a cube from a sphere, once he will recover his sight, will he be able to distinguish the cube and the sphere without touching them?[32]

Locke will reply by distinguishing ideas developed via a certain sense and those common to most senses: the former can be achieved automatically also by another sense. Many other philosophers, among which Berkeley, Voltaire, Buffon, Condillac in *Traité des sensations* (1754), Charles Bonnet (*Essai analytique sur les facultés de l'âme*) and Kant (in *Anthropologie*, 1798), will later reply negatively.

The same Molyneux will justify his negative answer with the claim that it is not possible to recognize "at first sight" the shape and the three-dimensionality of an object, in the same way in which they are experienced via the hand. The debate raised by the thought experiment is important, not only because of the confutation of innate ideas but also because of the relationship between hand and eye, which will become central in psychology and in the cognitive sciences later on.[33]

Berkeley who, as previously said, will also take a negative stand in the controversy, in the essay on the theory of vision (1709) wants to analyze the differences between sight and touch and verify if there are ideas in common between the two senses. Even by valuing—in contrast with Locke's position—touch over sight, he comes to the conclusion that the objects known via the two senses and the data provided are heterogeneous.[34]

---

[31] Descartes (2001).

[32] Locke included Molyneux's problem in the second edition (1694), of his *An Essay concerning Human Understanding*, II, 9 § 8: "Suppose a Man born blind, and now adult, and taught by his touch to distinguish between a Cube, and a Sphere of the same metal, and nighly of the same bigness, so as to tell, when he felt one and the other; which is the Cube, which the Sphere. Suppose then the Cube and Sphere placed on a Table, and the Blind Man to be made to see. Query, Whether by his sight, before he touched them, he could now distinguish, and tell, which is the Globe, which the Cube". See Morgan (1977).

[33] Gallagher (2005).

[34] "The extension, shapes, and motions perceived by sight are specifically distinct from the ideas of touch called by the same names; there is no such thing as one idea or kind of idea common to both senses". Berkeley (1709).

The idea of heterogeneity between the data provided by the two senses is dominant also during Enlightment among sensist and materialist philosophers, who, however, will argue, in an anti-cartesian manner, in favor of perceivable knowledge, attributing to touch—the "most philosophical" of the senses according to Diderot[35]—the primacy traditionally attributed to sight.

Also, Kant seems to value more touch when he distinguishes between "objective senses" (touch, sight, hearing) that contribute to the knowledge of objects thanks to an empirical intuition, and "subjective senses" (taste and smell), which don't provide knowledge, but enjoyment. However, even if we consider important, the data provided by touch, he judges the sense of sight as the most noble, because it is the one that is the most distant from touch; it has a wider perceptive sphere in space, and being less dependent from the organ's attachment, it is closer to the pure intuition.[36] In fact, even Hegel, in his *Esthetics*, believes that only the "theoretical senses" (sight and hearing) are suitable for artistic enjoyment, while sense of smell, taste, and touch capture the materiality as such and modify the object with which they relate.[37]

The analysis of Martin Heidegger on the hand's value is well known. Starting from the courses of the Forties,[38] he states that only the man, who, unlike the animal, can speak, can and must have the hand and, therefore, the gesture.[39] The most human gesture is, according to Heidegger, the hand writing (*Handschrift*), which protects and transmits the language. In his critique of the domination of technology, the philosopher highlights the dangers of typewriter, which would deny humans of "the dignity of the hand." The typographic mechanization, by subtracting the word to the hand, destroys its identity and unit, reducing it to a simple communication tool and hiding the personality of the one who writes.[40] However, it is important to notice that Heidegger does never use the plural "hands" but the singular "hand." His critique of the "optic paradigm" of western philosophy and his promotion of the hand are to be understood, therefore, in an ontological rather than

---

[35]Diderot (1751). Also Condillac (1754) values touch.

[36]"This sense (touch) is also the only one of *immediate* external perception and for this very reason it is also the most important and most reliably instructive, but nevertheless it is the coarsest, because the matter whose surface is to inform us about the shape of the object through touching must be solid". I. Kant (2006, 46–48).

[37]"Consequently the sensuous aspect of art is related only to the two theoretical senses of sight and hearing, while smell, taste, and touch remain excluded from the enjoyment of art. For smell, taste, and touch have to do with matter as such and its immediately sensible qualities—smell with material volatility in air, taste with the material liquefaction of objects, touch with warmth, cold, smoothness, etc. For this reason these senses cannot have to do with artistic objects, which are meant to maintain themselves in their real independence and allow of no *purely* sensuous relationship. What is agreeable for these senses is not the beauty of art". Hegel (1823).

[38]Heidegger (1998). This is the course run in Freiburg in the winter semester 1942/1943 and published in 1981.

[39]These theses will also be deepened in the courses Heidegger run in the winter semester 1951–52, published in 1954: (1976).

[40]See the comment to these Heideggerian theses by Derrida (1987).

epistemological perspective. In his interpretation, the "seeing," which has characterized the way of thinking from Greece onwards, means knowing and making the entity present to be used and owned. It is, therefore, very far from a contemplative dimension and hides a relationship with touch, or more precisely with hand, closer than it seems.[41] It is a relationship of domination, which foreshadows the will of power of technology: it is a seeing that is also a taking. In the context of Heidegger's critique to the technocratic paradigm and of his defense of humanism, we can better understand the meaning of the eulogy he does not to the hand, but to its essence. He considers the hand not as a prehensile organ—as the claw or the animal paw are—but as an expression of thought that opens itself to reality, thus receiving it. His hand is a disembodied hand, meaning it is as an opening of the thought that, through the word, shows and offers things and the world: it is the symbol of preserving and taking care rather than of dominating. Thanks to the phenomenological reflection on the corporeality, the issue of the function of the hand and, more in general, of the sensory perceptions is set within a unitary perspective, which sees the *Körper*—the physical body—and the *Leib*, the lived and sentient body as inseparable. According to this view, tactile sensations are particularly important: since they are localized, they refer both to the *Körper* and to the *Leib*. For example, when our hands are touching each other, we have both kinesthetic sensations, which are those referring to the hand when it is conceived objectively, and tactile sensations, through which we perceive the touched hand as a part that is inseparable from that sensory organ which is our living body, while the hand that touches is perceived as any external body.[42]

Merleau-Ponty's analysis shows the unity of sensory experiences, where perception is not understood as the sum of visual, tactile, and auditory data, but as the process through which these get integrated.[43] However, he also identifies the tangible with the visible: sight is the sense which is most capable of recognizing and imagining; still, the primacy of vision stands and it depends on the detached opening toward reality, without the appropriation provided by the hand, because with sight we "touch the world."[44]

Philosophy of corporeity, that also relies on Lacan's psychoanalytical investigation, has recently given value to the hand and touch in interpersonal relationships, which are essential for an acknowledgment of the other that is not cold and distant.[45]

---

[41]Petrosino (2006).
[42]Husserl (1913), Slatman (2005).
[43]*Phénoménologie de la perception* (1945).
[44]*Le visible et l'invisible* (1964), *L'oeil et l'esprit* (1961).
[45]Irigaray (2011).

## 4 Is the Hand Seeing or the Eye Touching? The Contribution of Experimental Psychology

The theory of *embodied cognition*, developed by the second-generation cognitive sciences, which overcomes the traditional Platonic–Cartesian dualism, gave back dignity to the body and to the sensorial experience. The data coming from developmental psychology and from neurosciences show that self-awareness is formed as one learns the distinction between one's own and someone else is thanks to tactile feeling. In fact, one of the first experiences of a newborn baby is the dual tactile feeling of the touch of his hand on his own face. It will be the investigations of neurophysiology and of psychology on the value and the importance of touch to provide to philosophy important elements for a wide perspective of the characteristics and the relevance of perception. While philosophy argues whether sight has only a theoretical function and touch only a practical one, modern psychology too goes through various phases with contrasting positions,[46] before reaching the conclusion that the hand has both functions: it is capable of perceptive actions with an epistemic goal and of practical actions with an instrumental goal.

Already in 1906, the biologist C. Sherrington had studied the so-called active touch, distinguishing it from the passive one to underline the conscious intention of fingertips in exploring objects.[47] David Katz, in his *Der Aufbau der Tastwelt* (1925),[48] reports the results of the experiments run starting from 1914 at the Psychological Institute of Göttingen, suggesting to get rid of the traditional division of the senses in "superior senses" (sight, hearing, taste, smell) and inferior ones (touch) and highlighting the importance of the latter as source of knowledge.

Observing finger movements, he derives a wide range of tactile feelings: of temperature, pressure, vibration, roughness. They show how fingers are a unitary tactile organ for consciousness that gives important information on the objects. Touching and feeling are not the same thing, as well as looking and seeing, hearing and listening.

Révèsz coined the term "haptic perception" or tactile-kinesthetic perception.[49] Studying the cognitive process of blind people, he states that the visual system is characterized by its simultaneity, while the haptic system has to gradually integrate the data at different stages, thanks to the explorative movements of the hand.

Erwin Straus,[50] in contrast with the Cartesian or deterministic definition of modern psychology, stresses the unity of the sensitive experience and argues with Von Senden,[51] who did not accept the idea of "tactile space" for people born blind,

---

[46]Piaget (1936), for example, believes that touch "teaches" sight in the perception of the third dimension; for Bruner (1959), children are able to visually recognize objects which are known only through other sensory means, like touch. Hatwell (1986).

[47]Sherrington (1906).

[48]Katz (1925).

[49]Révèsz (1950).

[50]Straus (1956).

[51]Senden (1960).

saying that they build the spatial scheme only on the basis of a series of temporal perceptions that convey the idea of extension. Straus, instead, believes that, if it is true that with touch one can only perceive a part of space, it is, however, thanks to these subsequent impressions that it is possible to get the idea of space.

Gibson offered two important contributions that allow a more complete reflection on the relationship between hand and sight: the distinction between active and passive touch[52] and that between sensation and perception. Active touch allows one to know the world since it does not consist in a simple stimulation of the hand or of the skin—such as being touched, but in the hand movements explicitly guided by the mind that is exploring (perception–action). With the second distinction, Gibson denies the classical theory of the heterogeneity of the data provided by sight and touch. Sensation is, in fact, the experience that follows the organ's stimulation, while perception is the information contained in the stimulation: different sensations can, therefore, integrate with each other to provide the same information.[53]

Accordingly, not only sight must not be "taught" by touch—as argued by Berkeley and Piaget—but both can contribute with similar perceptual mechanisms to the knowledge of objects and space.[54]

## 5 Conclusion

In conclusion, the analysis provided here shows how in the relationship between hand and sight the comparison between philosophy and experimental sciences helps to highlight on the one hand the complexity of the cognitive process, and on the other hand the unity between body and mind, showing that the opening of man toward the world is possible thanks to his essential embodied condition.

The study of interaction between touch and sight and, more generally, of the unity of sensitive experience and intellectual cognition is of fundamental importance also from the educational point of view. In fact, the so-called *digital generation* presents a capacity of attention guided mainly by the sense of sight and hearing, with a manual sensibility very much impoverished compared with past generations which had been in possession of a mainly tactile knowledge.

---

[52]"Active touch refers to what is ordinarily called *touching*. This ought to be distinguished from passive touch, or *being touched*. In one case the impression on the skin is brought about by the perceiver himself and in the other case by some outside agency". Gibson (1962).

[53]For example, the feelings of heat, the smell of something burnt and of smoke provide the perception of fire.

[54]On the other hand, the most recent studies on the blind show that they reach equivalent performances to sighted people in the acquisition of spatial data not just through sensory compensation, but through kinesthetic codification. A blind man uses tactile exploration better than a sighted man, but relies on his own body and on the codification of his own movements rather than on space. Hatwell (2003).

In those professions aimed at the bodily care of others, like medical or nursing service, where physical nearness and gestures have decisive importance, the exercise of sensibility, in particular, the tactile one, is extremely precious, because it supplies intelligence with that interconnection of analogies and intuitions which are necessary to be able to judge, to valuate, to compare.[55]

The tactile exploration and palpation carried out by medical personnel do not only imply a function related to the object or symptom, but also an interpersonal importance since they form a gesture of recognition able to treat the body of the patient like an organism where a person is made tangible. The *right of being touched* expresses an anthropological truth: the need of every human being, healthy or sick, to be protected and confirmed of his own identity and personal dignity, of which the body is an essential dimension.[56]

## References

Aristotle, *De Anima*.
Aristotle, *De partibus animalium*.
Aristotle, *De sensu* 1, 437 a.
Aristotle, *Etica Eudemia*.
Aristotle, *Historia animalium*.
Aristotle. (1960). *Metaphysics*. (W. D. Ross, Trans.). Oxford: Clarendon Press.
Berkeley, G. (1709). A New Theory of Vision and Other Select Philosophical Writings, introd. by A. Lindsay, J. M. Dent & Sons, London 1910.
Berninger, V. W., Abbott, R. D., Jones, J., Wolf, B. J., Gould, L., Anderson-Youngstrom, M., et al. (2006). Early development of language by hand: composing, reading, listening, and speaking connections; three letter-writing modes; and fast mapping in spelling. *Developmental Neuropsychology, 29*(1), 61–92.
Berteletti, I., & Booth, J. R. (2015). Perceiving fingers in single-digit arithmetic problems. *Frontiers in Psychology, 6*, 226.
Bruner J. S. (1959). Learning and Thinking. In *Harvard Educational Review*, 29 (3), 184–192.
Butterworth, B. (2005). The development of arithmetical abilities. *Journal Child Psychology and Psychiatry, 1*(46), 3–18.
Classen, C. (2012). *The Deepest Sense: A Cultural History of Touch*. Chicago: University of Illinois Press.
Condillac (1754). *Traité des sensations*. Londres-Paris: De Bure l'aîné.
Connally, K., & Elliott, J. (1972). The evolution and ontogeny of hand function. In N. Blurton Jones (Ed.). *Ethological Studies of Child Behavior*. Cambridge: Cambridge University Press.
Davidson, P. W. (1972). Haptic judgments of curvature by blind and sighted humans. *Journal of Experimental Psychology, 93*, 43–55.
Derrida, J. (1987). *Geschlechte II*: Heidegger's Hand, trans. John P. Leavey Jr. In J. Sallis (Ed.), *Deconstruction and Philosophy*. Chicago: University of Chicago Press.
Derrida, J. (2000). *Le toucher, Jean-Luc Nancy,* Paris: Galilée.
Descartes, R. (1637). *Discourse on Method, Optics, Geometry, and Meteorology. First Discourse. Optics*. Transl. P. J. Olscamp, Hackett Publishing Company, Indianapolis/Cambridge 2001: 65.

---

[55]Cfr. Hayes and Cox (1999), Edvardsson et al. (Edvardsson et al. 2003).
[56]Derrida (2000, p. 85).

Diderot, D. (1751). *Lettre sur les sourds et muets*. Paris: Jean-Baptiste-Claude Bauche.
Edvardsson, J. D., Sandman, P. O., Rasmussen, B. H., (2003). Meanings of giving touch in the care of older patients: becoming a valuable person and professional. *Journal of Clinical Nursing*, *12*, 601–609.
Gallagher, S. (2005). *How the Body Shapes the Mind*. Oxford: Clarendon Press.
Gibson, J. J. (1962). Observations on active touch. *Psychological Review, 69*(6), 477–491.
Gibson, J. J. (1966). *The Senses Considered as Perceptual Systems*. Boston: Houghton Mifflin Company.
Gracia-Bafalluy, M., & Noël, M.-P. (2008). Does finger training increase young children's numerical performance? *Cortex, 44*, 368–375.
Hall, E. T. (1966). *The Hidden Dimension*. Garden City, N.Y.: Doubleday.
Hatwell, Y. (1986). *Toucher l'espace. La main et la perception tactile de l'espace*. Lille: Presses Universitaires de Lille.
Hatwell, Y. (2003). *Psychologie cognitive de la cecité precoce*. Paris: Dunod.
Hayes, J., & Cox, C. (1999). The experience of therapeutic touch from a nursing perspective. *British Journal of Nursing*, *8*(18), 1249–1254.
Hegel, G. W. F. (1823).Aesthetics. Letters on Fine Art, trans. T. M. Knox, 2 vols. Clarendon: Oxford 1975, 1: 2.
Heidegger, M. (1976). What Is Called Thinking?, trans. J. Glenn Gray. New York: HarperCollins.
Heidegger, M. (1998). Parmenides, trans, A. Schuwer & R. Rojcewicz. Bloomington: Indiana University Press.
Husserl, E. (1913). Ideas Pertaining to a Pure Phenomenology and to a Phenomenological Philosophy. Second Book: Studies in the Phenomenology of Constitution, 1989. trans. R. Rojcewicz and Schuwer, A. Dordrecht: Kluwer, II § 36.
Irigaray, L. (2011). Perhaps cultivating touch can still save us. *Substance*, 40(126), n. 3 130–140.
Kant, I. (2006). Anthropology from a Pragmatic Point of View (1800), §§ 16–19, (Ed.) R. B. Louden, Cambridge University Press: Cambridge.
Katz, D. Der. (1925). *Aufbau der Tastwelt*. Leipzig: Johann Ambrosius Barth.
Kress, G. (2003). *Literacy in the new media age*. London: Routledge.
Le Breton, D. (2006). *La saveur du monde. Une anthropologie des sens*. Paris: Métailié.
Lewin, K. (1934). Der Richtungsbegriff in der Psychologie. Der spezielle und allgemeine Hodologische Raum. *Psychologische Forschungen*, 19, 249–299.
Mangen, A. & Velay, J.-L. (2010). Digitizing Literacy: Reflections on the Haptics of Writing. In Mehrdad Hosseini Zadeh (Ed.). *Advances in Haptics*, InTech. Available from: http://www.intechopen.com/books/advances-in-haptics/digitizing-literacy-reflections-on-the-haptics-of-writing .
Merleau-Ponty, M. (1945). *Phénoménologie de la perception*. Paris: Gallimard.
Merleau-Ponty, M. (1961). *L'oeil et l'esprit*. Paris: Gallimard.
Merleau-Ponty, M. (1964). *Le visible et l'invisible*. Paris: Gallimard.
Montessori, M. (1949). *The Absorbent Mind*. Oxford: Clio Press.
Montessori, M. (2006). *The Secret of Childhood (1936)*. New Delhi: Orient Longman Limited.
Morgan, M. J. (1977). *Molyneux's Question: Vision, Touch and the Philosophy of Perception*. Cambridge: Cambridge University Press.
Napier, J. R. (1956). The prehensile movements of the human hand. *The Journal of Bone & Joint Surgery*, 38-B(4) 902–913
Newman, S. D., & Soylu, F. (2013). The impact of finger counting habits on arithmetic in adults and children. *Psychology Research, 78,* 549–556.
Pallasmaa, J. (2009). *The Thinking Hand: Existential and Embodied Wisdom in Architecture*. Chichester: John Wiley & Sons.
Paterson, M. (2007). *The Senses of Touch. Haptics Affects and Technologies*. Berg: Oxford-New York.
Petrosino, S. (2006). Visione. In *Enciclopedia Filosofica Bompiani*, vol. 12, Milano: Bompiani, 12183–12188.

Piaget J. (1936). *La naissance de l'intelligence chez l'enfant*, Neuchâtel; Paris: Delachaux et Niestlé [*The origins of intelligence in the child*, trans. M. Cook. London: Routledge & Kegan Paul, 1953].

Plato, *Teeteto*.

Révèsz, G. (1950). Psicology and Art of Blind, London and New York: Longmans, Green & co. Available online: https://archive.org/stream/psychologyartofb00gr#page/10/mode/2up .

Senden von, M. (1932). Space and Sight: The Perception of Space and Shape in the Congenitally Blind Before and After operation, trans. P. Heath (1960), New York: Free Press of Glencoe.

Sennett, R. (2008). *The Craftsman*, Yale University Press, New Heaven & London, p. 161.

Sherrington, C. (1906). *The Integrative Action of the Nervous System*. New York: Scribner's Sons.

Slatman, J. (2005). The sense of life: husserl and merleau-ponty on touching and being touched, *Chiasmi International*, VII, 305–325.

Straus, E. (1956). *The Primary World of Senses: A Vindication of Sensory Experience*. New York-London: MacMillan.

Tallis, R. (2003). *The Hand: A Philosophical Inquiry in Human Being*. Edinburgh: Edinburgh University Press.

# *Ready-to-Hand* in Heidegger. Philosophy as an Everyday Understanding of the World and the Question Concerning Technology

José Manuel Chillón

**Abstract** With this paper, I will try to investigate why *ready-to-hand* is one of the first philosophical findings on which all subsequent philosophical experience must be built. Therefore, I will establish a first approach to the concept *ready-to-hand* in *Being and Time*. Secondly, I will study the matter in the context of the phenomenological research about life considered an original problem for philosophy. Finally, we will discuss how this affects the relationship between man and the *handiness world* in terms of care (Sorge) in connection with technique and modern technology.

**Keywords** Ready to hand · Heidegger · Technology · Phenomenology · Facticity

## 1 *Ready-to-Hand* and Overcoming a Theoretical View of Philosophy

The third chapter of the first section of the first part of *Being and Time* deals with the world as a constitutive moment of Dasein's existential structure. In the second chapter, Heidegger observed that *be-in-the-world* refers to a non-theoretical and objective relationship (or mathematical one as Descartes understood) with the world in which, above all, Dasein inhabits. Inhabiting the world is our first and fundamental experience of the world as an area of belonging which is already given. Heidegger's perspective in *Being and Time* is related to his initial research: his focus on practical-instrumental life and, therefore, on human action (Heidegger 1985); on a certain way of understanding human identity from the existence which solders the being of being which is Dasein to the being which makes beings be. He also points out the ontological difference which highlights the peculiarity of Dasein's existence. This finding overcomes the ontical perspective of metaphysics in which there is not a difference between Dasein's projective existence and other

J.M. Chillón (✉)
Department of Philosophy, University of Valladolid, Valladolid, Spain
e-mail: josemanuel@fyl.uva.es

innerworldly beings' mere presence. This analysis of Dasein involves the discovery of the worldliness of the world, that is, the discovery of the world as a phenomenon. The world is made up of entities that are things and precisely, because of this, Heidegger explains that "being a thing" (*Vorhandenheit*) is the problem which reminds us of the classical problem of substantiality: the way of being of things on which everything is based. The world is not a mere innerworldly being, but rather the condition of giving of these innerworldly beings, writes Heidegger (1996, 67). That the world is a character of Dasein does not rule out that the world can be analysed as of the innerworldly beings. The being of the world, *worldliness*, cannot be captured by an ontic study because it cannot notice the fundamentality of Dasein's structure as *being-in-the-world*. The most immediate world which Dasein *has to be in* is the *surrounding world*: our everyday world (Heidegger 1996, 62). For Dasein, *being-in-the-world* means *the association in the world* with the innerworldly beings which we encounter. This *association* is dispersed in manifold ways of *taking care of* as Heidegger explains (1996, 68). And this is, in my opinion, the key to the analysis of worldliness in Heidegger: *being-in-the-world* is always *taking care of* the things that are in the world. *Taking care of* innerworldly beings does not involve a theoretical knowledge of the world, because these beings are things that are used or that are produced. This is made possible by phenomenological access to these beings, i.e. by overcoming the theoretical view that actually conceals the phenomenon of *taking care of*.

The beings we encounter are useful. The world is a context of meaning or a connection of instrumental references. Heidegger refers to these references as the relationship between useful things whereby these useful things do not appear alone but as part of a complete articulated structure. And this structure, invisible to theoretical understanding, is somehow perceptible for practical cognition. *Aufleuchten* is the verb chosen by Heidegger to define this prior understanding which, incidentally, goes unnoticed by daily treatment. When and how can we capture the essence of this pragmatic structure? Exactly when this useful thing stops working, either because it has been damaged, or there is some impediment or disturbance.

Things *go out to meet* human beings as tools. *Going out to meet* is Heidegger's way to translate the *giving of things* as a relevant phenomenological concept. Useful things Heidegger writes—never "*are*" because they are always "*in order to*". They always have a structure of referentiality. "Strictly speaking, there 'is' no such thing as a useful thing", explains Heidegger (1996, 64). The way of being of useful things (in which *useful* reveals itself by itself) is *handiness*. And this way of being (which the Greeks called *ta pragmata* without knowing the strict character of *pragmata*) cannot be understood from a mere theoretical perspective. But this does not mean that association which makes use of things is blind. "It has its own way of seeing which guides our operation and gives them their specific thingly qualitiy. Our association with useful things is subordinate to the manifold references of the *in-order-to*" Heidegger points out (1996, 65). Its own way of seeing is called *circumspection*. This peculiar way of doing philosophy, having to adjust to the referentiality of *in order to,* takes the form of *circumspection*. So, Heidegger says, the theoretical perspective is a way of seeing which does not need a method. *Ready-to-hand* is not a subjective determination of things; "handiness is the ontological categorical definition of beings as they are in themselves". (Heidegger 1996, 67). *Ready-to-hand* is the ontological status of things in the world from the perspective of the ontological difference that distinguishes between the way of being of the being that is Dasein and the other worldly beings that are there to use, to grasp, to touch. The structure of *ready-to-hand*, as a useful thing, is made by references, by the *in order to* which determines the plexus of meanings. These many references also determine a familiarity with innerworldly beings that Dasein *takes care of*. Everything at hand is mutually referred to each other. Or, what is the same, everything at hand is always for something and that *in order to* of each being is the meaning. The practical agent is the one which gives meaning to things in the context of the significance of the world. The world is, then, the place where the possibilities of the beings exist while these possibilities are linked to the possibilities of a practical agent, the possibilities of Dasein. The significance of the world, with which Dasein is already familiar, is the condition of possibility whereby Dasein may come into contact with the world and interpret it in terms of meaning. Obviously, overcoming an exclusively theoretical foundation involves considering these meanings not from a purely logical perspective and makes us refer rather to schemes of action or schemes of useful treatment.

## 2 *Ready-to-Hand* and the Phenomenological Problem of Philosophy

Philosophy seems to have a strong intellectual aspect. By dint of insisting on the theoretical perspective of philosophy, western tradition has lost sight of the only philosophically important issue: the life which we are already living. Finding oneself living (which means understanding that one exists) is the same as discovering oneself *being-in-the-world*. The first notion that human beings have about the world in which they live is that this world is *handiness* and therefore, they can touch it, they can use it. Anthropogenesis is on the side of Heidegger: the increase in cranial capacity comes after achievements such as our opposable thumb for grabbing, our upright posture and the consequent release of the lower extremities. It is clear that man's first encounter with life is not theoretical, but practical (praxis), through the action which connects us with the world: touching. Philosophy must start with this finding, points Heidegger. And this reveals what the theoretical interpretation of Husserl's phenomenology had obviated: the facticity of human existence. The discovery of *being-in-the-world* is made only from a practical perspective: the hermeneutical transformation of phenomenology that occurred in Heidegger. His capital intuition is consistent with this hermeneutical transformation of phenomenology: it is not an intuition of the *eidos* of things in a profound theoretical exercise still anchored in the epistemic subject-object structure (despite Husserl's attempts to free himself of this). But it is recognised that the same thing that must be accessible to philosophy is life (Heidegger 1985). And life refers to *being* always *in the world*. The way of being in the world, which is what is already given, invites philosophy to move from theory to understanding (*Verstehen*). And, what is understanding? Give philosophical relevance to Dasein's being in the world. A world that now is a plexus, a context of meanings and worldliness as being the things of the world. Well, we can understand that the world as a whole, which cannot be touched, which cannot be grasped at once, is composed of things that are, above all, useful tools which we can touch, use, discard, transform, etc. And this is a hermeneutical-existential way to give meaning to the world in which I am, either my world, the surrounding world, or the world of us. Western philosophy from Plato, as it gave preference to *Vorhandenheit*, was not aware of the worldliness of the world because of the reduction of the world to something merely *in sight*, to something merely present, to a merely theoretical object. *Present-at-hand* (*Vorhandenheit*) and *ready-to-hand* (*Zuhandenheit*) are the ways in which beings (that do not have the way of being of Dasein) may occur. *Zuhandenheit* refers to immediate, practical and hermeneutical relationship between Dasein and beings before any theoretical understanding. *Vorhandenheit*, for its part, explains the way of being of things that are there to be contemplated and not only to be used. This last perspective has constituted the whole tradition of Western metaphysics which has understood being as always being and that Heidegger firstly calls *metaphysics of presence* and then, *ontotheological project of metaphysics*. Phenomenology, to be true to itself, must ignore its scientific turn (as if it could be a phenomenological science of immediate pure data which is not given anywhere), to access *something* which simply was given before. Heidegger provisionally delimits this with the concept of *life itself*. "In an epistemological way, there is no absolutely pure experience" explains Heidegger (1993, 134). Obviously life itself, because it is our life, cannot be studied from a distance as if it were a separate being.[1] "Our theme is Dasein in its being-there for a while at the particular time" says Heidegger (1999, 37). If phenomenology must be the science of the *forehaving*, this original meaning must be like our practical everyday knowledge before theoretical knowledge, like something in a field before objectivity. And this is Dasein's *being-in-the-world*: factical life. This is facticity as a philosophical theme: the fact of living and having to do so. Somehow, we always appear in life. And life is the world in which we live, warns Heidegger (1993, 34). The subject-object scheme (that favours theoretical knowledge) leaves the initial and fundamental factical situation, in which things are understandable in the first instance, on account of the link between Dasein and the world. So, his facticity and, therefore, his *being-in-the-world* is, above all, hermeneutical. Hermeneutics aims to understand *the being* at all times. Hermeneutics also aims to clarify the meaning of having to *be-in-the-world* with others and with innerworldly beings. The task of philosophy is to be *phänomenologische Hermeneutik der Faktizität,* points Heidegger (1999, 15). Any experience which occurs is already in the context of a reality which has been

---

[1] If so, we would reduce the phenomenon itself of being which is always an attempt of thinking. Heidegger (1999, 4).

pre-interpreted somehow (*Ausgelegtheit*), in a horizon of meaning in which world and Dasein are linked. This is the decisive hermeneutical transformation of phenomenology. The *already being* in the world is the fertile source that should guide the steps of philosophical phenomenology.

However, because the primary relationship between Dasein's life and the world is understanding (*Verstehen*), this facticity can be approached through hermeneutics. Hermeneutics is, therefore, the phenomenological way of approaching, of accessing, of questioning and of explaining facticity. The *Faktizität* is about Dasein's *own being* and, therefore, its identity includes aspects such as contingency, finitude and historicity.

This factical perspective implies Dasein's essential way of being defined in terms of *care*. Care must be seen as a fundamental phenomenon of Dasein's *being there*, explains Heidegger (1999, 104). Young Heidegger had called *Sorge* to the intense *being engaged* in something (*Aus sein auf etwas*). This care (that replaces intentionality as a fundamentally theoretical relationship between Dasein and the world with *how to behave* towards something), sees the world from the *directions of care* that Heidegger called *worlds of care*. In his early work, Heidegger distinguishes *surrounding world* (*Um-Welt*), *with-world* (*Mit-Welt*) and *self-world* (*Selbst-Welt*). Care is the essence of Dasein's being. So, *being-in-the-world's* existence does not mean to be like the other things which are innerworldly beings. *Being-in-the-world* is about serving, manufacturing, addressing business, taking possession of something, preventing or protecting from damage, writes Heidegger (1999, 102).

Having the world *ready-to-hand* and having to *take care of,* it is an essential task of Dasein. In my opinion, technique and, in a sense, technology could be considered a contemporary way of understanding how man uses things which are *ready-to-hand.* However, has the evolution of technique and modern technology respected the essence of Dasein, which is taking care of the world as a way of letting the world be? Or, has man degraded the essence of *handiness* to such a point that he has also essentially degenerated himself?

## 3 *Ready-to-Hand* Technique and Technology: Between Danger and Saving

Perhaps, the first thing to think about is the difference between technique and technology because it seems that we are using these words indistinctly. Alfredo Marcos explains that technique refers primarily to a skill or practical process, a know-how and its products, be they objects or actions. Technology can, therefore, mean two things: (i) either a scientific knowledge of *tekhne* or (ii) a *tekhne* accompanied by science, that is, a know-how accompanied by a know-why. We shall normally understand technology to have this second meaning, as *tekhne* accompanied by or derived from science. In fact, although the etymological distinction is clear, in practice, the words *technique/technic(s)* and *technology* share a common semantic field distributed differently by different languages. (Marcos 2010, 565).²

In fact, the text we take as reference, *The question concerning technology,* distinguishes between the essence of technique and its general notion, technology as a set of artefacts and the heideggerian theses about the technology of the future. The more separate the semantic fields of both words become, the less will be understood the fertility of heideggerian analysis whose originality has precisely to do with technology as the supreme expression of technique which implies, from the beginning, a way of getting to the object that is genuine of metaphysics. The German word that Heidegger uses is *Technik.* This word is used for both *technique* and *technology* although specialists have generally reserved "technique" for the old issues related to the use of instruments and "technology" for the more scientific applications. Not surprisingly, the English versions of the heideggerian lessons on technique have been translated by the word *technology.* But, let us go to our subject: Why is it important for man to ask about technique?

Insomuch as all questions are already a search, all questions mark a starting point on the path of thought. So, the question about the essence of technique must be asked from *thoughtful thinking*, from thinking that meditates. In the same way that the question of metaphysics is only possible from a way thinking which marks a distance from traditional metaphysical thinking, the question of technique is claiming a space reserved for the criticism of a technical civilisation that has obviated thinking about technique. When Heidegger writes *Being and Time*, the author is not even critical of modern technology. What appears in this book and in other previous writings, mainly those derived from the influence of his study of Aristotelian physis, are the basics of this subsequent criticism of technique. However, according to Volpi (1988), there is a structural reason which explains why the Heidegger of *Being and Time* could not criticise technique and modern technology as he would later. The fundamental attitude of Dasein in its confrontation with things is the attitude of *taking care of* that is characterised in *Being and Time* in a positive way. But then, Heidegger warns that modern technology has degraded the meaning of that *taking care of* by absolutising use. I want to try to understand how some of these normative aspects of dealing with the world reappear by returning to *Being and Time*.

*Being in the world* is understood as *inhabiting the world.*³ Inhabiting is part of a plexus of meaning, and it is starting to establish a relationship with a context that does not belong to us because it was there before us. Therefore, the starting point of

---

²In fact, it is a question of explaining the increasingly diffuse boundaries between the natural and the artificial. *"The development of technology has entailed drastic changes specialy in the field of new materials, energy resources and the processing and communication of information. We live sorrounded by artefacts and we have altered our relationship with the natural world".* Marcos 2010, 567.

³According to Peter Sloterdijk, a treatise about being and space appears under the title *Being and Time*. The old verb *innan* reveals the key of the existential analysis of spatiality. *Being in the world*

any philosophical reflection should be precisely the practical-instrumental experience. All these aspects which are relevant for understanding technology were already laid down in 1927 in *Sein und Zeit*. Technique should be an extension of how man uses the useful things (the handiness world) that Dasein finds in front of him. It should be guided by the care, the essence of Dasein, whereby man considers useful things useful. Technique and modern technology have also altered the relationship of man with his understanding of the world so this practical connection becomes a submissive relationship. By not letting the world be world, by not letting a useful thing be only useful, the identity of things is deformed because things are transformed from mere accessories to conditions of being that consequently degrade human identity. Heidegger explains that in its familiarity with things, Dasein finds its meaning. Dasein's meaning is found both in its being and in its being able to be in relation to its *being in the world*, explains Heidegger (1996). So modern technology is consummate metaphysics because it takes the dissolution of the ontological difference to its final consequences: human beings become useful. This is the most deplorable inversión. If this relationship between human beings and useful things is altered, the fundamental experience whereby Dasein finds meaning in the world is degraded.

Technology involves *an unconcealment* (bring-there-front). Not surprisingly, since the *Nicomachean Ethics* of Aristotle (Aristotle 1999, 1139b 18), technical knowledge is one of the *ways of being in the truth*.[4] This is what leads Heidegger (1977a b) to discover that technique is not about a mere human instrumentum, but a way of being which refers to man and, at the same time, a way of revealing what there is. The question of the essence of technique and modern technology cannot be disentangled from the fundamental question that every man should ask: the question of being. Not noticing the *oblivion of being* (which vertebrated modern metaphysics) was nowadays reflected in technology, *the consummate metaphysics* (*Gestell*). Consummate metaphysics understood as the most overwhelming expression of forgetting the *oblivion of being*. What does mean *oblivion of being?* According to Volpi, since the early 1930s, the idea that metaphysics, in order to complete its task of capturing, subjugating and subduing *physis*, has to draw on the modern technology that is actually the most expeditious determination of the Greek conviction of understanding *being* as a constant presence, matures in Heidegger (obviously influenced by E. Jünger).[5] However, modern technology has its origins

---

(Footnote 3 continued)

means *innen*, *being inside of the world* in a transitive sense: inhabitting the world in its openning up to agreements made before.

[4]Heidegger considers technique a fundamental aspect of Dasein's care, that is, a way of *living in truth*. This is one of the ways in which Dasein is already in the pre-understanding of being, therefore Heidegger finds the basics of what would be his genuine proposal: the hermeneutics of facticity.

[5]Most of these reflections on technology are found in *Grundfragen der Philosophie. Ausgewählte "Probleme" der "Logik"*, 1937–1938. The edition of Friedrich-Wilhelm von Herrmann. V. Klostermann, Frankfurt a. M., 1984, GA 45.

in the beginning of Greek thinking and in the ambiguity of its understanding technique as an unveiling of *physis* and as a way of capturing it. In our world, technology shows a peculiar characteristic of western thinking, appearing as a form of *total mobilisation*, in Jünger's words (Kitler 2008).[6]

We live in technology and not just with it. But our need to question the essence of technology disappeared when man stopped questioning meaning, that is, the question concerning not only what life means, but what technology essentially means for human life. And this is the basic thesis of Heidegger: both excess modernisation and escaping the necessity of thinking are involved. And the serious problem in *The question concerning technology* is precisely that the unceasing progress of technology causes the unfortunate circumstances of man. Heidegger is aware of the void of our time. The void exists because the only important thing is to make everything work. Technology penetrates deeply into the nihilistic character of our time, thus implementing the time of *the will to will* that *lording* of things, which homogenises every being as well as dominates everything.[7] In fact, although the *oblivion of being* cannot be imputed to philosophical thought, modern technology has some responsibility for preventing the possibility of man ever being conscious of the *oblivion of being*.

Well, the key elements of technology are what Heidegger calls *setting-upon* or enframing (*Ersetzbarkeit*), namely the need that technology has to view nature as a single and giant service station, and man, who believed himself to be the lord of the earth, as a reserve stock of what has been called human material.[8] This is the highest degree of instrumentalisation in which every being becomes essentially replaceable in a generalised exchange in which everything can be changed for anything. Where this essential *enframing* dominates, which is also anonymous because no one is responsible for what happens, where man is seen as a reserve store of any kind for any intended improvement of man himself, here lies the greatest danger. "Meanwhile man, precisely as the one so threatened, exalts himself to the posture of lord of the earth". Heidegger (1977a, b, 17).[9]

This immersion in modern technology (that goes unnoticed for the human being) decreases man's chances of considering other ways of *being-in-the-world* that are

---

[6]Kitler was strucked by the number of parallels in both the argumets and the terminology of these texts: *The Arbeit* (1932) of Ernst Jünger and *The question concerning technology* of Heidegger.

[7]In "*The age of world picture*" Heidegger explains that such uniformity becomes the surest instrument of the complete technical dominion over the earth. Subjectivity's modern freedom dissolves itself in objectivity.

[8]Unfortunately, Heidegger's argument is developed at such a high level of abstraction he literally cannot discriminate between electricity and atom bombs, agricultural techniques and the Holocaust. All are merely different expressions of the identical enframing, which we are called to transcend through the recovery of a deeper relation to being. In Heidegger, Feenberg explains, is difficult to see in what relation would consist beyond a mere change of attitude. (Feenberg 2000, 297).

[9]Modern science, Agazzi awards (1998, 80), was born of the pretension towards manipulating being that is the core of technology, and this implies, as a consequence, an attitude of *violence* that underlies technology. Heidegger was the initiator of this doctrine, which has found many followers in the present intelectual climate.

not domination and subjugation. The techno-scientific mentality has taken hold of our time and has finally besieged man himself. But where danger lies, a saving power also grows, explains Heidegger collecting the verses of Hölderlin. The emergence of what can save us is only possible if man continues to question, the essence of technology with regard to the same mystery of his being a man.

Our technological understanding of reality is based on this preconception of the entities as nonsense in a nihilism that considers them as accounting products ready to be modified and used. Is not this current criticism of the environmental devastation or the constant biogenetic obsessions or experimentation with transgenics based on this subjugation of everything to Gestell in that *technological ontotheology* of our time marked by the *will to power*? Perhaps, with Thomson (2009, 160–161), we must recognise that the value of Heidegger's analysis is not so great for threats of danger as for the redemptive promise of thinking about nothing in which the technified world plunges us as an opportunity to think differently exceeding the ontic reductionism of the technique as Gestell. In my opinion, in Heidegger, a moderate stance on technology halfway between naive optimism and fatalistic pessimism is being born. In short, it would try to value the technique from the difference between overcoming by overcoming and overcoming by human well-being.

Under the heideggerian analysis, the nietzschean difference between negative nihilism and nihilism as the starting point for a new sense is latent. The first is the nihilism that serves as an atmosphere to the technological civilization that, in turn, feeds and increases the density of this nihilism in which the void that our present lives happens. The second is the nihilism that arises when everything is meaningless and, then, there is the opportunity to distance oneself from the technology, to ask about it and so, to save the human being.

So, in Heidegger's works, there is no demonisation of technology[10] precisely because this danger becomes, in turn, an opportunity for renewing man's commitment with truth just at this particular moment of maximum rupture. This moment is now. This moment is our time. Heidegger calls *thinking* this insubordination to technology, this critical distance with technology, this need to return again and again to the essence of technology. And now is the right time to think. So, *thoughtful thinking* involves distancing ourselves from beings in order not to be trapped by them and not becoming a piece of a system that works. *Thinking* the distance is realising the difference that must always be between beings and Dasein; a distance which is a critical capability which enables *thinking* about being. Then, the trouble is not that the world has been completely modernised, but that man does

---

[10]Indeed, Heidegger always refused to put his speech in the context of the pessimistic speeches of those who, at the time, referred to *the decline of the west*. Heidegger takes over the nietzschean sentence: desert grows, but the fact that the fundamental problem is we do not yet think, seems to bring a ray of light to the horizon of desertification.

not know how to face this problem by *thinking*. *Thinking*, although unable to redeem us from the pernicious fate of a technological humanity, at least alerts man of the quicksand enveloping him and of the time that will destroy him just as it is about to deify him.

So, what should humanity's position regarding technology be? This question is all the more important because the answer will affect how man should *be in front of* beings. In *Being and Time,* Heidegger (1996, 78) explains that "the way of being of *things-at-hand* is relevance". Relevance means ontically "to let *things at hand* be in such and such a way in factical taking care of things, to let them be as they are and in order that they be such".

I think that one of the important keys to Heidegger's analysis of modern technology appears in *Being and Time*: take distance from things. That is, let things be what they are without confusing them with ourselves. Taking distance from things is to respect what they are without trying to change their ontological status. Heidegger writes that letting things be does not mean producing the being of something but "discovering something in each case as being, in its *being-at-hand*". To this regard, in §23 of *Being and Time*, Heidegger analyses the experience of the treatment of Dasein not in terms of construction but of discovery.

We should differentiate between ontic relevance which explains that relevance is the condition of the possibility of being in front of *things-at-hand*, and ontological relevance in connection with giving freedom to everything at hand. The expression *bewenden lassen* means letting something be or letting it remain as it is. Heidegger meanwhile uses *bewenden lassen* to indicate that the way in which *things-at-hand* are presented is interrelated with the letting be of Dasein. In fact, the first feature of the treatment of useful things is *Entfernen*, i.e. bringing *ready-to-hand* into the scope of Dasein, letting it come into its presence by bringing distance near. This distance is not the spatial distance which refers to availability of useful when it is necessary, depending on the course of action in which Dasein is immersed. Heidegger defines *distance* as a closeness that must never disregard the space that should always exist between entities and Dasein. Dasein, Heidegger explained, experiences the remoteness of handiness with respect to Dasein itself. A distance that Dasein can never cross over even though Dasein has an essential tendency to closeness.

Thinking of modern technology (in which the object is product of the *will to power* and in which man, as a subject, knows that he is trapped by the object) reveals the model of what we can summarise as Heidegger's philosophical legacy: the intimate belonging together between Desein and being. Heidegger calls this *Ereignis*: that is, the rise of a new dimension that is emerging (beyond all metaphysical tradition).

Technical objects should be used properly but they must remain free, Heidegger writes in *Serenity*. We can say "yes" to the necessary use of technique but we can also say "no" and we can refuse to use them exclusively, to bowing down, confusing and eventually degrading our essence". Serenity (*Gelassenheit*) is a particular state of mind (*Befindlichkeit*) that evokes *the mystery of being* in which context must be understood the issue of technology. Serenity is the attitude that can best describe the relationship between Desein and the world in terms of technology. Therefore, serenity becomes a critical attitude that enables us to discover the truth of being by respecting the difference between Dasein and being; a respect that modern technology had forgotten. Serenity could be a possible answer to the question that we did above about our position concerning technology. Serenity is the attitude with which man has to be in front of beings, and it is the attitude which reveals a new way of relating freely to things that are *ready-to-hand*.

## 4 Conclusions

Perhaps the first criticism is to admit that Heidegger, in the first third of the s. XX, could not think of the dangers of an unsuspected technological civilization: computer science, robotics, nuclear physics...[11] But this is not our position. Technique and technology, although one refers to the practical application of knowledge and the other to the technique linked to science, both in Heidegger's position, would fall under that set of strategies with which Dasein, at the expense of wanting to live better, can betray its essence, can degrade itself ontologically. In what sense? To the extent that it is a set of forms implemented to try to exceed human finitude, or at least to hide from contemporary man his dependence on being. This critique of technique, which can be extended as a warning to technology is the thesis behind its critique of humanism as the philosophical and cultural version of a whole western tradition which, by placing man at the centre of the world, understands everything else (and also all the others) as submitted to man. Hence, Heidegger's criticism of technology can not be made simply in terms of instrumental rationality but in the context of a critique of underlying metaphysics. And this is the decisively important thing that, otherwise, will go unnoticed: technology is no neutral. Our way of thinking of technology is revealing our understanding of human condition.

In my opinion, and in an aspect that I can only point to now, transhumanism and posthumanism are versions of this exacerbated humanism whose threat Heidegger spoke of.[12] It is, then, to think again what it means to care for the things that are *ready-to-hand* without this natural use accruing in abuse that alters the being of things. It is, again, to see what it means to take care of oneself without this care accruing in an improvement of life that alters human condition, that is, personal dignity and freedom. This is nowadays *the theme of our time*.

---

[11]This was the Bruno Latour's thesis about Heidegger's conception of technology. Latour (1999) considers that Heidegger's writtings on technology are fictitious, antiquated and pesimistic.

[12]Against what Sloterdijk says (2009).

# References

Agazzi, E. (1998). From technique to technology: The role of modern science. *Techné: Research in Philosophy and Technology, 4*(2), 80–85.

Aristotle (1999). Nicomachean ethics. Kitchener: Batoche Books.

Feenberg, A. (2000). From essentialism to constructivism. Philosophy of technology at the crossroad. In: E. Higgs, A. Light & D. Strong (Eds.), *Technology and the good life* (pp. 294–312). London: Chicago University Press.

Heidegger, M. (1977a). *The question concerning technology and other essays.* New York & London: Garland Publishing.

Heidegger, M. (1977b). The age of the world picture. In *The question concerning technology and other essays* (pp. 115–154). New York & London: Garland Publishing.

Heidegger, M. (1993). *Grundprobleme der phänemenologie.* Frankfurt am Main: Vittorio Klostermann.

Heidegger, M. (1985). *Phänomenologische interpretationen zu Aristoteles. Einführung in die phänomenologische Forschun.* Vittorio Klostermann: Frankfurt am Main.

Heidegger, M. (1994). Basic questions of philosophy. Selected "problems" of "logic". Bloomington: Indiana University Press.

Heidegger, M. (1999). *Ontology. The hermeneutics of facticity.* Bloomington & Indianapolis: Indiana University Press.

Heidegger, M. (1996). *Being and time.* New York: State University of New York Press.

Kitler, W. (2008). From gestalt to ge-stel. Martin Heidegger reads Ernst Jünger. *Cultural Critique,* 79–97.

Latour, B. (1999). *Pandora's hope. Essays on the reality of science studies.* Cambridge: Harward University Press.

Marcos, A. (2010). Tact-glossary: Technology. *La Clinica Terapeutica (Clin Ter), 161*(6), 565–567.

Sloterdijk, P. (2009). Rules for the human zoo. A response to the *Letter on humanism. Environment and planning. Society and Space. 27,* 12–28.

Thomson, I. (2009). Understanding technology ontotheologically. In J. Olsen, E. Selinger, & S. Riis (Eds.), *New waves in philosophy of technology* (pp. 146–168). Hampshire: Palgrave Macmillan.

Volpi, F. (1988). Dasein comme praxis: L'assimilation et la radicalisation heideggerienne de la philosophie pratique d'Aristote. In J. Taminiaux (Ed.), *Heidegger et l'ideé de la phénoménologie* (pp. 1–43). Dordrecht: Kluwer Academic Publisher.

# The Therapeutic Hand

### Xavier Escribano and Albert Pérez-Bellmunt

**Abstract** The purpose of our work is, first of all, to rekindle a sense of wonderment as the basic starting point for a philosophical reflection on the hand, initially focusing on the special versatility, ductility and protean character of the human hand. It is then a question of examining and reassessing its inimitable practical dimension, especially in the field of health, as an essential part of the art of healing and caring. Following the inspiration of classical works concerned with the relationship between doctor and patient, one might argue that in medical practice the hand maintains a balance between specialisation and multifunctionality. In the field of health care, despite a necessary reduction of initial possibilities and an inevitable specialisation of its tasks, the hand maintains its characteristic multifunctionality. In fact, the therapeutic hand is, all at the same time, instrumental, cognitive and pathic. Of all these features, the one that appears as the least technified or least specialised, that is the pathic or affective hand, is, nevertheless, the bedrock with respect to the others. The pathic or affective dimension is the very basis of the therapeutic hand, since it is this that drives it to act for the good of the patient, and it is this dimension that in some way must direct and involve any operation of the therapeutic hand, whose ultimate purpose is the care and healing of the human person.

**Keywords** Therapeutic hand · Instrumental hand · Noetic hand · Pathic hand

---

X. Escribano (✉)
Philosophical Anthropology at the Faculty of Humanities and the Faculty of Medicine and Health Sciences, Universitat Internacional de Catalunya, Barcelona, Spain
e-mail: xescriba@uic.es

A. Pérez-Bellmunt
Anatomy at the Faculty of Medicine and Health Sciences, Universitat Internacional de Catalunya, Barcelona, Spain

## 1 Introduction: A Paradox About the Hand

In the cultural history of the West, we might note the striking paradox of a special theoretical or speculative interest in the human hand which is accompanied at the same time by a practical contempt for all manual occupations. For one thing, there is the insistence on the eminent dignity of the hand in contrast to other parts of the body, as an organ which is appropriate—and to an extent analogical—to the qualities of openness and universality of the mind. In contrast, any kind of manual labour is regarded as something demeaning, an obligation undertaken out of need and which involves a difficult and thankless contact with matter, as opposed to the exercise of purely intellectual tasks, befitting more the spirit and exclusive to it. In short, the hand is praised, but its activity is not held in high esteem. This, then, is the paradox.

An example is Aristotle who, in many passages, favours the human hand over any other animal limb, calling it «the tool of tools» (Aristóteles 2003), and he draws an analogy between the soul and the hand, the two constituting a common space. However, in describing the type of life which is noblest and happiest for humans, he sees this in purely theoretical activity, rather than in any manual task (Aristóteles 2014). Similarly, in the Middle Ages, we find the theologian and philosopher Thomas Aquinas intervening with the *Quaestio disputata*, *De opere manuali*, in the debate that takes place at the University of Paris concerning the relationship between manual and intellectual work, which in the culture and mentality of that time are in dissociation with each other, if not in total opposition (Peig 2007). Again in the thirteenth century, the physician and surgeon Lanfranc, in his *Chirurgia Magna*, calls for greater respect by the medical establishment for the manual skill which surgical practice implies. Medical knowledge and the manual practice of surgery appear in this academic context as dissociated disciplines or even dissociated professions, because surgery is not purely intellectual and involves technical operations and a direct contact with the living, palpitating and blood-bearing matter of the human body. In this same field, at the gates of the twentieth century, the famous French clinician Dieulafoy also dissociates the surgeon's hand and the intellectual activity of the doctor (Pera 2003).

In our own day, with the development of cybernetics and robotics, we see attempts to emulate the action of the hands mechanically, perhaps in the hope that certain artefacts can better it in precision and reliability and eventually overtake it so that as an instrument it would—paradoxically—be found wanting next to its own creations. For the moment though, as Roesch says, we are still far from being able to produce a robot with motor skills comparable to human ones. «Robotic hands are probably the best example of this failure: we build ever-sophisticated robotic devices, but we cannot figure out how to use them in the type of seamless interactions with our world that they are meant to enact» (Roesch 2013, p. 412).

Considering the above, the purpose of our work is, first of all, to rekindle a sense of wonderment as the basic starting point for a philosophical reflection on the hand, initially focusing on the special versatility, ductility and protean character of the human hand. It is then a question of examining and reassessing its inimitable practical dimension, especially in the field of health, as an essential part of the art of healing and caring. We have in mind especially the view of Cristóbal Pera in his original and pioneer *Diccionario filosófico de la cirugía*, where he comments on this when he looks at the relation between the surgeon and the patient, though his observations could be extended and applied to any therapeutic relationship in general: «In current medicine, there is a predominant trend for manual contact between the surgeon and his patient to be lost. This has been an increasing tendency in social medicine, reflecting organisational defects rather than being intentional, one which limits to a ludicrous extent the communication between doctor and patient and sometimes obliges him, and at other times perversely accustoms him, to dispense with a manual exploration which he tries to replace in vain with a series of X-rays or analyses» (Pera 2003, p. 235).

For this reason, at the present moment, it seems opportune to make this reflection, which seeks to rediscover and revalue the versatility and multifunctionality of the human hand in the therapeutic field. This initial versatility will materialise later, throughout our exposition of three especially prominent functions: instrumental, noetic and pathic. Finally, the need for a balance and a mutual interweaving between this plurality of dimensions will be shown, emphasising the bedrock character of the pathic or affective hand.

## 2 The Extraordinary Versatility of the Hand

On 17 October 1938, Paul Valéry, the celebrated French poet and essayist, was invited to give the opening speech at a Congress of Surgery that was being held in Paris. In this text, aptly entitled *An Address to Surgeons* and notable for its substance and originality, we find some passages that could be considered a true eulogy to the human hand, a multifaceted part of the body which—in the words of the poet —«strikes and blesses, gives and receives, feeds, takes an oath, measures, reads for the blind, speaks for the dumb, is extended to a friend, is raised against the adversary and becomes a hammer, a pair of pliers, an alphabet» (Valéry 1993, p. 178).

It is not surprising to find that he includes a digression about the possibilities of the hand in a speech to surgeons, since when we look at the etymological derivation of the word 'surgery' we see that it comes from the Greek word *cheirourgía*, made up of *cheír* (hand) and *érgon* (work). What greatly surprised Valéry was that there was not, in his own day, any *Treatise on the Hand*, that is, there existed no comprehensive study of the very many capabilities of this «prodigious machine that combines the most nuanced sensitivity with the most unbound of forces». However, Valéry, with such a suggestive evocation and such intelligent descriptions of this «organ of the possible», this «organ of positive certainty» and this «universal agent» which realises our dreams, was wrong in thinking this. In fact, more than a hundred years before, in 1833 to be precise, Sir Charles Bell, a Scottish surgeon and anatomist, had published his magnificent study *The Hand, Its Mechanism and Vital*

*Endowments as Evincing Design*, which remains an essential reference work to this day and a true classic of comparative anatomy.

In this work, Bell attributed a privileged place in nature to the hand and, comparing it thoroughly with the structure of the limbs of other animals, observed in it a wonderful architecture that led him to think—hence the title of his work—that there was evidence of intelligent design. The hand is in fact, as recent research also shows, an anatomical region with a bone structure that allows one the possibility of opposing the thumb to the other fingers with a system of muscles and ligaments that facilitates the forming and creating of all kinds of digital pincers. All these are fundamental actions for the evolution of our species and are made possible by the peculiar and unique shape of the human hand (Day et Napier 1963; Tocheri et al. 2008).

The lyrical evocations of Valéry and the anatomical descriptions of Bell both manifest their admiration for an organ whose ductility and versatility are unique. Since ancient times, there has been awareness of the famous debate between Anaxagoras and Aristotle on whether humans are rational because we have hands or whether we have hands because we are rational. That discussion, at the dawn of philosophical thought, is not too far from the focus of a recent publication by the Croatian teacher and researcher Zdravko Radman, entitled *The Hand, an Organ of the Mind* (2013a), in which specialists from different fields show to what extent the movement and actions of the hands are critical in shaping mental functions related to knowledge and action. However, it goes beyond establishing a mere causal relationship between the faculties and manual abilities, something about which both earliest Antiquity and the latest research are in agreement and emphasise the very same anthropologically relevant fact—the striking analogy between the qualities of the mind and the almost unlimited versatility of its main projection in the human body. If intelligence seeks to understand everything, the hand seeks to dominate everything; if intelligence can conceive of any device, only the hand is able to materialise it (Barbotin 1970). The liberation of the hands from the locomotion of our body means that they become suitable instruments for action in the world about us and grant a naked and vulnerable animal the vital protection of clothing, fire, shelter, tools or symbols. The hand which holds, receives, gives, blesses or promises is the foremost of instruments enabling us, rational beings of flesh and blood, to dwell in the world as if it were our home.

There is too Galen, the great physician and anatomist of Antiquity, who dedicated Book I of his *De usu partium* to the study of the hand. In that treatise, the doctor from Pergamon, following in the footsteps of his teachers Hippocrates and Aristotle, emphatically stresses the mutuality of, and the parallelism between, reasoning and manual dexterity. In the same way that man has not been endowed with rigid instincts which direct his conduct in one single direction, neither has he been graced with hyperspecialised bodily organs for one particular environment. Rather «in exchange for the nakedness of his body, he received hands, and in exchange for the lack of skills of his soul he received reason (…) Hence man, the only living being who has in his soul the most excellent ability, has in his body, according to this logic, the most excellent of instruments» (Galen 2010, pp. 94–95).

Compared to the claw, the hoof or the fin, extremities of animals which are highly specialised and tailored to a particular environment, the human hand, with its lack of initial specialisation, appears to be an organ which is tragically indefinite and completely useless for survival. But this lack of specialisation is compensated by its close relationship with the brain, owing to countless nerve connections and its ample cortical representation. Thus, the hand has a great functional plasticity and is —thanks to its powers of tactile discrimination—one of the best illustrations of the sense of touch and the ability to act. This is not just a result of the development of techniques through which it makes and handles all kinds of instruments that act as an extension to itself in a thousand different ways, but because the hand itself, which has not come into the world naturally endowed for any specific ability, remains ever open to learning all of them: from the prodigious dance of the pianist's fingers on the keyboard to the pinpoint accuracy when making an incision with a surgical scalpel and from the sophisticated symbolism of hand gestures in Kathakali theatre or Odissi dance in India to the frenzy of signs with fingers and hands of the stock traders in the Chicago Stock Exchange, for example—the object of study of the choreographer and researcher in cognitive sciences Ivar Hagendoorn.[1]

To perform this whole sequence of operations that its protean quality allows it to do, the hand has, more than anything, to be a free organ. Having been freed from the functions of locomotion, which in our bipedal system is relegated to the lower extremities, the upper extremities can concentrate on other tasks. They can be sent out, as it were, to explore the world, to play with its materials or textures and transform it in a thousand different ways. Citing an example from a well-known classical text, one could say that, unlike what happens in Ovid's *Metamorphoses*, where the transformations of human bodies by the gods express the subjection of human life to forces that man cannot control, the successive metamorphoses of the hand express the plurality of dimensions that result from actions determined by our own volition. This was highlighted by the celebrated Scottish anatomist Sir Charles Bell when he established a close correlation between the designs of the will and movements of the hand: «The human hand is so beautifully formed, it has so fine a sensitivity, the sensibility governs its motions so correctly, every effort of the will is answered so instantly, as if the hand itself were the seat of that will» (Bell 1979, pp. 13–14). If the human body, in the experience we each have of it, is the «organ of the will and bearer of free movement» (Husserl 1952, § 38), one could say that, of all the members that make up our bodily structure, this assertion is especially true in the case of the hands, with their agility and unparalleled skill. Conversely, this strong association between functional versatility and the freedom of action is a telling reminder of the problem posed by functional deficiencies that affect the mobility of the fingers or hands, since they restrict or significantly impair the ability of an individual to engage in activity (Cole 2013).

---

[1]Cf. http://www.ivarhagendoorn.com/video/the-fisher-account.

## 3 The Therapeutic Hand: Instrumental, Noetic and Pathic

So far, we have insisted on the special design of the hand, which bestows on it a versatility and multifunctionality that has been a source of admiration since ancient times. Moreover, there is no doubt that many activities and professions require a reduction of its initial versatility, a concentration on some possible features of this multifaceted and multi-instrumental organ. For example, the sculptor's hand, the hand of the pianist, the karateka's hand or the mechanic's hand imply, each in its field, a restriction and a specialisation of the extensive options that the hand possesses. As for medical practice, the thesis that we put forward here is that, despite the relative specialisation that certain techniques may require, the therapeutic hand keeps alive—within its specificity—an ample range of functions that relate to the plurality of levels implicit in the practice of medicine itself or related healthcare professions (scientific, technical, humanistic levels).

Following the inspiration of classical works concerned with the relationship between doctor and patient, such as that of Laín Entralgo (1964) or more recent ones like those of Van Manen (1999) and Pera (2003), one might argue that in medical practice the hand maintains a balance between specialisation and multi-functionality. That is to say, in one sense we have the hand which moves, massages and manipulates, and which operates, transforms and modifies, using one instrument or another, according to a function that could be considered technical or instrumental (*technical or instrumental hand*); in another sense, we have the hand that palpates, strikes or squeezes, undertaking an exploration corresponding to a cognitive function (*noetic or cognitive hand*); and finally, we have the hand offered receptively to a patient as a gesture of recognition (or respect), which supports, consoles, encourages or accompanies, serving a function which is social, emotional and interpersonal (*pathic or affective hand*).

### 3.1 The Technical or Instrumental Hand

Freed from the task of locomotion and released once and for all from its contact with the ground, the hand is able to come into its own in the world. «I experience my hand», says Erwin Strauss, «as an organ in relation to the world» (Strauss 1952, p. 545). The hand's relation to the world can take many forms: exploring a surface is not the same as handling a tool or caressing a face. The instrumental hand, which is the foundation of material culture, moves out into the world; that is, it intervenes in it and transforms it. The most basic and primordial form of culture, a necessary complement for a naked and defenceless animal, devoid of natural specialisations, consists precisely in the transformation and exploitation of the conditions of nature through the work and activity of man himself (Gehlen 1987), and for this, the close collaboration of free hands and intelligence is needed.

The hand shows a special adeptness in its relation with the objects of the world since it combines both ductility and firmness. That is, while the hand can surround and adapt to the shape of the object, as if it has no form of its own, it can also be closed tightly in the act of gripping or taking hold of an object. Before and during prehension, the hand «preshapes and reshapes itself» (Mattens 2013, p. 169). Like Rilke's angels in *Duino's Elegies*, we can pause to look with admiration at the familiarity with which the hands engage in a dialogue with matter, in its various textures and varied forms, making use of a kind of inherent knowledge. It seems as if any object, before entering into a more direct and close relationship with our personal and intimate space, goes, so to speak, through a process of mediation, negotiation and manual examination that appraises its characteristics and approves or disapproves: «The hands facilitate the evaluation of objects for potential manipulation» (Gallagher 2013, p. 214). Because of the hands, a pragmatic area of possible activities or manual action, of the handling or use of objects, is projected around the body, which is what Herbert Mead in 1938 called a «manipulatory area» (Gallagher 2013, p. 214). Consequently, the perception of the world immediately surrounding us is determined by the aspect of its eventual manageability, manipulability, tangibility, etc.: «we begin to understand situations in terms of what we have 'at hand' prior to any manifest purposefulness or deliberation. Space becomes 'hand centred'» (Radman 2013b, p. 378).

The instrumental hand, in the old formulation of Aristotle, already quoted, is *the instrument of instruments*. Indeed, the instrumental hand not only intervenes itself, directly, in dealing with objects and in the transformation of the world, but also gives rise to all kinds of utensils and tools, which diversify, specialise and amplify its own capacities. It could be said that the hand has continuation in the tool, expanding corporeal spatiality and the corporeal schema, transforming itself and adopting this temporary extension of itself as an amplification of its own being. In spite of this intimate relationship between the hand and the tools it employs, the human being is free to use one tool or another since the hand is not bound to any of its instrumental extensions.

Even so, the hand is not indifferent to the use of certain instruments, especially in the case of very habitual or intensive manual activity. For this reason, we may find that the hands are a mirror of the profession practised, or the action that they have been performing. The hands of the sculptor or potter differ from the hands of a writer, and the imprint left by years of trauma surgery is not the same as if they had been engaged in neurosurgery. The modelling or deformation of the hand resulting from the instruments with which it works, however, can also be a symptom of its possible dehumanisation, of its reduction to the condition of being a mere implement. The hand, in order to be fully human, must maintain its freedom at all times from the instruments it uses; otherwise, it becomes instrumentalised or transformed into an object. The freedom of the hand with respect to its instruments safeguards its versatility and the universality of its function. Freedom in the use of one tool or another is, in essence, a consequence of the fact that the production of these tools is also free and not a constraint of nature. Indeed, as Hans Jonas points out, in the human being, the production of tools is free, as it does not arise from any organic function and it is not subject to any biological programming (Jonas 1998). The freedom resulting from their invention and the freedom in their use are the two faces of the same phenomenon.

A tool serves as an intermediary between the executing body organ (hand) and the extracorporeal object of the action. In turn, tool mediation or interpolation can be multiplied. Mediation in ascending powers (Jonas 1998) can be continued indefinitely, but all of them refer back ultimately to the hand. In a technological civilisation such as ours, the hypertrophy of instrumental mediation can be an inconvenience in that it implies a growing distance between the agent and the product and between the hand that acts and the object of its intervention. In the field of health, a good example is robotic surgery equipment (e.g. the *Da Vinci* surgical system) in which, from a console, the surgeon controls three or four interactive robotic arms, at the end of which different instruments (scalpels, scissors, separators, etc.) are attached. This system permits the optimisation of the range of action of the human hand, the reduction of any possible tremor and the perfection of all the surgeon's movements, although it still requires, for the moment, a learning process by the healthcare professional who uses them.

This use of technology and robotics in current medicine makes surgical intervention less and less dependent on the «art of the hands of the surgeon» and more on the technical capacity of the instruments used. Consequently, in the case of the therapeutic hand to which we are referring, technological mediation presents this ambivalence: firstly, it offers the advantages of mathematical precision, exact calculation and movement which is controlled and exact to the millimetre; secondly, however, there is no doubt that at the same time the technology is rather like a dividing partition in the doctor–patient relationship that sometimes prevents or hinders direct treatment. Touching or manipulating a body is also a way to recognise it and to welcome it, as it were, in its humanity.

The adeptness or technical skill of the hands is essential in medical care: we can imagine the hands of a nurse inserting a catheter, the hands of a physiotherapist performing manipulative therapy on the vertebrae, the hands of a dentist extracting a tooth, the hands of a doctor carrying out an endoscopy. In all these operations, both the success of the treatment or intervention and the attempt to keep pain to a minimum depend directly on a skilful deployment of the hands and the correct handling of the instruments utilised.

In its therapeutic function the instrumental hand always affects another human body, thus having the life of another person in its power. This is particularly true in the case of the surgical hand (the surgeon's hands), which is referred to by Cristóbal Pera: «What does the Surgeon do with his hands? In the first place, he develops a technique in the operative field, a kind of know-how, in which his hands are the instruments of instruments: he cuts, he dissects, he binds together, he sutures. All these actions he accomplishes with a personal rhythm, which attempts to be, at one and the same time, energetic and caressing, marking the movements that strive to be beautiful through eurythmy» (Pera 2003, p. 234). In the hands of the surgeon, the relationship between life and death is felt with a special intensity and with a characteristic ambiguity to which Valéry is especially sensitive in his address to the surgeons: the surgical hand, in order to heal, holds within it a weapon, and in order to heal, it wounds. This closeness of the surgical act to the most transcendental realities of living and dying has a symbolic and ritual dimension that emphasises the liturgy of asepsis, attire and gestures in an operating theatre (Valéry 1997).

The relevance or protagonism of the instrumental hand in the therapeutic domain, that is the hand that intervenes by manipulating and penetrating the human body, is related to the concept of medicine that guides it. It can be said, as Hans Georg Gadamer has pointed out, that the essence of the art of curing does not consist in producing something new that did not exist previously, as is characteristic of all other techniques, but that «it belongs to the essence of the art of healing that its ability to produce is an ability to re-produce and re-establish something» (Gadamer 1996, p. 32). According to Gadamer, the true work of medical art consists of «an attempt to restore an equilibrium that has been disturbed» (Gadamer 1996, p. 36). Consequently, «Among all the sciences concerned with nature the science of medicine is the one which can never be understood entirely as a technology, precisely because it invariably experiences its own abilities and skills simply as a restoration of what belongs to nature» (Gadamer 1996, p. 39). However, we can ask ourselves how far removed this conception and ideal of medicine, which connects with its more classic inspiration in the Greek origins of science, is from the medicine of the Western world, with the molecular medicine, genetic engineering and cosmetic surgery (not reconstructive surgery) of today.

We may ask ourselves whether the hand that intervenes technically to restore a natural balance is being replaced progressively by a hand that intervenes to create or produce a new nature. If that is the case we are moving from technology in the service of medicine, to medicine as a sophisticated technology, that uses the organic as the raw material of its activities. Cyborg fashion and the fantasies of transhumanism suggest this. We might also ask ourselves whether in 'non-invasive' surgery, which attempts to alter and traumatise the human body as little as possible, we are not in fact witnessing a far more profound invasion—an excess of intervention, we might say. Along with the invasion in the organic body, it is the invasion in the metaphysical body, that is in the natural definition of the body. Finally, we might also ask ourselves whether the technological extension of the potency of the hand in its therapeutic rôle does not cause us to believe increasingly in the myth of an omnipotent medicine, capable of curing everything, because it possesses the most intimate secrets of nature. With this, the instrumental hand emerges as if a magical, thaumaturgical hand.

## 3.2 The Noetic or Cognitive Hand

Touch and physical contact are fundamental to care efforts, and therapists usually touch their patients, consciously or unconsciously, in their interactions with them

(Routasalo and Isola 1996). Different studies have focused on the different uses of touch and their effects (Routasalo 1999). But one of the most outstanding and specific characteristics of the sense of touch is its reversibility or bipolarity. Whenever I touch something foreign to myself, any external object, at the same time I feel the part of my body that is touching it. The tactile properties inherent in the object (roughness, hardness, concavity, etc.) correspond to tactile sensations in the subject (the sensation of touch in the tip of the fingers, tingling, heat, pain, pressure, tension, etc.). It is not possible to touch without being touched. Despite the importance for one's own body that Husserl attaches to tactility (*Taktualität*) (Husserl 1952, p. 150), or the central role played by the double-sided nature of *touchant/touché* in the phenomenological and ontological conception of corporeality, for example in Merleau-Ponty (1945, p. 108; 1964, p. 172 et seq.), the fact is that this aspect is traditionally presented as secondary to a sensation directed primarily and specifically at the object. It seems as if the important and fundamental aspect of touch was the ability to recognise the shape and texture of objects.

Nevertheless, as Filip Mattens points out, the fact that touch is useful for an awareness of sensations on the surface of the body itself does not seem to be something of lesser import to the interests of a living organism. The sensitivity of the skin, which covers the whole body of the animal, rather than for touching, serves to know where it is touched, which is evidently of great importance in protecting its integrity and for its survival. Thus, the sense of touch is not necessarily and essentially «object directed» (Mattens 2013, p. 164).

This reflection helps us to better understand the exceptionality of the hand as an active organ of touch. In fact, the hand stands out because of its ability to feel and so directly recognise certain physical qualities of material reality in all its concreteness and inalienable presence: «[t]he touch of the hand lets us explore the materiality of the world around us» (Van Manen 1999: p. 26). If the human body is a system of systems dedicated to the inspection of the world, then the cognitive hand, or manual touch, is remarkable within the body for the proximity, depth and intimacy that define its interaction with the peripersonal world that is within its reach, the «manipulatory area» discussed above. On the other hand, although the noetic or cognitive hand, which we might also call the exploratory hand, is notable as an active organ of touch, against a backdrop of tactility which occurs all over the body, this active function does not make it alien to the characteristic reversibility of touch. In fact, manually exploring an object is a way of directing our attention to something, with a level of personal involvement that surpasses that of any sense that operates at a distance, as is the case of sight or hearing: «[w]hile touching, man is in a 'sensitive' way personally involved with the materiality of things, which is hidden for the distance-senses» (Buytendijk 1970, p. 113). This is an objective inspection with a special level of subjective involvement, due to the reversibility or bipolarity which is characteristic of touch. As Hans Jonas says in his article *The Nobility of Sight*, touch does not enjoy vision's causal indifference (Jonas 1954). It is not possible to touch without acting on the object and without the object acting on oneself. It is not possible to feel tactilely without interacting with the tangible. To put it even more simply and clearly, it is not possible to touch without modifying the object touched and, in turn, be modified by the object touched. This fact will be of particular importance, and we shall emphasise below.

In terms of evolution, manual exploration has the advantage of releasing the front part of the head, the face, which in the animal remains as a snout or muzzle, from the task of carrying out an evaluation of the more immediate materiality of the object. Thus, the human face—liberated, we could say, thanks to the previous freeing up of the hands—can become a vector of expression and a personal face, an unmistakable sign of identity (Polo 1991). However, despite the relative independence that some parts of the body seem to grant to others (the feet releases the hands from the task of locomotion, and the hands release the face from the tasks of immediate sensory inspection, etc.), the activity of each component takes place in a joint, coordinated fashion. Gnostic touch acts in close collaboration with the gnostic eye and with the gnostic mind, forming a complex and unitary system that contributes to the performance of the act. As Merleau-Ponty points out in various places, the human body, in its diversity of parts, acts synergistically and can orient all of itself to make a single gesture in which it concentrates all its activity (Merleau-Ponty 1945).

In the therapeutic context, which is of particular interest to us in this study, the noetic or cognitive hand could be defined as a «data collecting instrument manipulated in a diagnostic manner" (Van Manen 1999, p. 23). Some features of the hand make it especially suitable for this task: «The human hand is marvellously equipped to be receptive to different types of sensations. Medical and nursing handbooks provide a great deal of detail about the practice of palpation and the sensory discriminations detected through the use of the hand. Because of its anatomical structure the hand possesses regional sensitivities and degrees of receptivity to different type of sensations. The finger pads are more sensitive to tactile discriminations for detecting moisture, contour, consistency and mobility. Finger tips are especially suited to explore tiny skin lesions. And the dorsal surface and the ulnar edge of the hand and fingers are most sensitive to variations of temperature. Vibratory impulses are best detected with the palmar surface, or ball, of the hand» (Van Manen 1999, p. 22). As a result, it could be said that the diagnostic touch is a «specialisation of the more general cognitive and probing aspects of touch» (Van Manen 1999, p. 26), which has, in all medical traditions, led to touching and feeling being considered as an additional tool in medical diagnosis (Elkiss and Jerome 2012): a tool whose use does not just depend on the hand's own capacities, but that requires a process of learning and can only be fully developed through the experience and skill of the therapist. In this way, the palpatory diagnosis goes beyond a technique and becomes an art. An art has led many scientific studies to attempt to establish to what extent it is reliable or valid, with varying results (Cooper et al. 2013; Fingleton et al. 2014).

Numerous authors coincide in highlighting the importance of movement in active manual touch, the intimate interpenetration of sensorium and motorium: «It is not the resting finger that feels, but the actively moved digits. Tactile impressions result from motion» (Strauss 1952, p. 546). The hand collects impressions moving along the surface of the object examined, passing along it as if it were some unknown territory, one which is yet to be discovered, pressing down and releasing the pressure, sliding along its contours and edges. The intertwining of the active and passive aspects of touch is emphasised especially by this close alliance between the motor dimension and the sensory dimension of touch (Radman 2013b, p. 193). The hand approaches and moves away from the object touched. That movement «continuously renovates the contact with the things, and … grants these a pronounced characteristic of reality» (Buytendijk 1970, p. 112). The flexibility and motility of the hand (Mattens 2013) make this indeed possible. In its act of exploration, the touching hand receives feedback from the touched object (Radman 2013a, b). This feedback from the initial contacts directs or guides the nature of further exploratory movements in a process that is circular or dialogic: «The movements bring about the perceptions, and these, again, the movements» Buytendijk (1970, p. 104).

Unlike sight, touch perceives the density of things, or, to put it another way, the resistance of things (Jonas 1954). That resistance is experienced as a force of opposition to my movement and to my progression. Bumping against or colliding with another provides me with the tangible certainty of his or her reality, in a very different way to sight or hearing. For this reason, as already mentioned at the outset, Valéry maintains that the hand is the «organ of positive certainty». Touching is the most reliable way of checking the reality of the other. The tactile sensation is accompanied by the impression of the resistance of the other, which in his or her reality opposes my own reality. The recognition of the reality of the other (Radman 2013b, p. 131) is an acknowledgement of one's own reality and of one's limits in relation to other realities: «Touching is existing in one's own boundaries (…) one becomes aware of oneself together with that with is touched» (Buytendijk 1970, p. 101). In the same vein as this author—and in keeping with what Merleau-Ponty would affirm—it could be said that all the senses are a means of communication, but each of them communicates in a different way. One of the more specific peculiarities of touch is to check the resistance and, consequently, the reality of the tangible. Now, paradoxically, in its ability to engage me and offer me resistance, this tangible reality is confirming, in turn, that it is part of my reality, that it is with me, that it is embedded in the same fabric of real things that constitutes the weave of my being and, consequently, that I can enter into communication or into communion with it: «because the feeling (or touching) human being finds the felt object as immediately present with himself (…) touch has, in human relationship, a fundamental spiritual meaning and represents communion in its human sense, too» (Buytendijk 1970, p. 102).

The fact that touching is a special way of communicating, and that this communication can become a form of communion, an *être a deux*, in the words of Minkowski (1999, p. 182), makes the vast difference between touching an inert body and, as is the case of medical practice, touching a human body especially relevant. There is no doubt that the human body can be touched as a purely objective body, in its merely physical materiality, as with an inspection which is naturalistic or scientific in purpose, or as a personal body (Laín 1964, p. 339). In the palpation or probing of the doctor, we find that both an objectifying and cognitive act and an interpersonal communicative act are in some way intertwined (Laín 1964, p. 341). Van Manen, who addresses this question carefully, distinguishes between a gnostic and probing 'touch of palpation' and a 'pathic touch of support'. He assigns one or the other kind of manual touch in accordance with the medical carer it predominantly corresponds to: according to Van Manen, on the one hand, we would have the medical gnostic hand and, on the other hand, the pathic nursing hand. For this author, while the «specialist physician's hand is primarily the gnostic hand, a knowing intellectual hand, a hand of science, as it diagnostically examines the human body for signs of trouble, or as it surgically operates on the body to remove a feared tumour», the nurse's hand would be «a healing hand, a caring hand which does not only touch the physical body, it also touches the self, the whole embodied person» (Van Manen 1999, p. 32). However, we might ask ourselves, is the therapeutic hand, both gnostic and pathic, not independent of the professional status of the health worker?

Although Van Manen's somewhat dual approach is initially clear, it does not stand up to scrutiny. Not only because, as Van Manen himself acknowledges, there is in the nurse's hand a gnostic–pathic ambiguity, which occurs in many healthcare situations and actually gives it a double function, but also since the doctor's hand has this simultaneous gnostic–pathic function and would continue to have it even though an objectivist, scientific spirit would roundly reject the notion that there was any emotional or affective implication. It cannot be denied that the intention guiding the action of the one who explores manually is a decisive, critically important one, as regards what kind of touch is being used, whether it is gnostic or pathic, but it is the very nature of touching, whatever the purpose or intention that is behind the action, which makes it difficult, indeed impossible, to distinguish with total clarity the gnostic aspects from the pathic aspects of the touch.

It has been said at the beginning of this section that unlike the senses that act at a distance, touch implies a personal commitment in the materiality of things. Touch, as Buytendijk puts it, quoting Lavelle (*La dialectique du monde sensible,* 1921), is charged with a «caractère d'émotivité charnelle et de vie», or as he says, quoting Nogué (*Esquisse d'un système des qualités sensibles*, 1943): «Le toucher transmet de la chair à la chair la palpitation de la vie la communion avec une destinée étrangère» (Buytendijk 1970, p. 115). Despite the fact that the intrinsic character of touching, i.e. immediacy, reversibility or bipolarity, transitivity of contact, is not essentially connected with the realm of desires or intentions, every manual exploration—by its very nature—goes beyond the limits of neutrality and detachment within which I might want to confine it.

Just as the practice of medicine conditioned by the designs of a merely instrumental hand seems bound to degenerate into a technology which leads to an artificial nature, the predominance of a purely diagnostic or cognitive hand would perhaps disregard the interpersonal aspect of the therapeutic relationship and would reduce the patient to an observable reality, without bringing to bear the inevitably affective and pathological nature of what, literally, the doctor would have in his own hands. In addition, the diagnostic or cognitive hand cannot avoid the spectre of a possible replacement—something already now in effect—of a manual examination by other diagnostic methods that imply a greater distance between doctor and patient. That is why, in addition to the instrumental hand and the cognitive hand, we must now deal with the pathic or affective hand, not only as an intrinsic aspect of manual touch, but also as one with an undeniable therapeutic meaning.

## 3.3 The Pathic or Affective Hand

Touch and physical contact have been part of the healing process in many civilisations and cultures over the centuries. Touching as therapy is one of the oldest forms of treatment in the world, first described in China during the second-century BC (Arnould-Taylor 1991). The gnostic touch of palpation and the pathic touch of support do not impose absolute and perfectly clear-cut boundaries, although both uses can be distinguished by the situation in which they occur, the intention behind each act of touching and even the way it is consented to or accepted by the person receiving some form of treatment. It has already been pointed out above that the ambiguity between the gnostic and the pathic is due to the very nature of the act of touching, so that in the end it is insuppressible. Though the palpation of a human body may be neutral, distant and objective in a medical and diagnostic examination, it never ceases to be an intrusion (benevolent and well-intentioned, in this case) into the personal and intimate space of another human body. Hence, the discomfort that some people experience from the simple fact of being touched by strange hands is understandable.

The «intersubjective touch» is, in any situation, the most basic form of social interaction and human communication (Farmer and Tsakiris 2013, p. 103). When, in a situation of visible suffering, the intention of the pathic touch is to provide support, comfort, company, reassurance, understanding, etc., it is necessary to take into account aspects such as the age, gender or culture of the person receiving the care and attention. There has even been talk of 'contact cultures' (e.g. Arabs, Latin Americans, southern Europeans) and 'noncontact cultures' (e.g. North Americans, Asians, northern Europeans). The same gesture can be understood or interpreted in very different ways, depending on the cultural codes, the beliefs or the personal circumstances of the individual in question.

Max Van Manen, in considering the care that nursing professionals offer their patients, distinguishes personal pathic touch from private pathic touch. Indeed, the caress of support or comfort of the professional caregiver does not have the same meaning as the loving or intimate caress of the lover (Van Manen 1999). The personal pathic touch, though invisible or little noticed in nursing care—since it has not as yet been sufficiently considered in nursing protocols—has a purpose or function intrinsic to the profession (reassurance, support, comfort, etc.). In the case of the elderly, it is of special importance. And in some cases, when other forms of expression are no longer possible, it can become the essential means of communicating with the patient, as Marie de Hennezel has shown in her experience in a palliative care unit with patients in a terminal situation: «Throughout my relationship with so many patients who were in distress and locked in the suffering of their ruined bodies, I have developed a means of communication through touch, a technique of touching the person which allows the person to feel whole and fully alive. It is as if the patient's sore skin needed to be coated with a second, more subtle and ethereal skin. A second psychic skin, a skin of the soul. Sometimes there is nothing that can replace the contact of one hand, because it truly establishes communication with the other. Now that I hold Louis's wan face in my hands, and his features are relaxed and his skin becomes infused with heat, I feel genuinely in contact with him. We have not said anything intelligible, but we are united.» (Hennezel 1996, pp. 191–192)

From an empirical point of view, and in addition to what has been stated above, certain scientific theories have sought to indicate the benefits with which *touch* provides the patient. Uvnäs-Mobertg and Petersson (2010) suggest that it may be due to the generation of the hormone oxytocin by our brain when we are touched. Oxytocin is commonly known to be one of the hormones responsible for happiness or well-being, and one of its many functions is the forging of bonds of affection. This could explain why patients undergoing chemotherapy have fewer bouts of nausea when they receive a massage (Billhult et al. 2007), the glucose reduction observed in diabetic patients after a tactile treatment or the importance of touch in the patient–therapist relationship (McCorkle and Pasacreta 2001).

The common feature shared by the gnostic touch and the personal pathic touch, as has been described here, is its professional component, the fact that it is employed in a context with healing and care as the purpose. Now, while the first tends to separate the person from his body, which it attempts to consider solely and exclusively as a physical reality, the second tends to reintegrate the person with his body, since the act of giving comfort or support, although carried out by something so apparently physical as a caressing hand, is not directed at the physical body of the person, but at the person himself in his bodily and spiritual totality: «this is a healing hand, a caring hand which does not only touch the physical body, it also touches the self, the whole embodied person» (Van Manen 1999, p. 33). The pathic personal touch does not remain on the fringes. It is not limited to discovering a symptom or a case of a general pathology, but rather touches the person in his or her essential core, and for that reason establishes—directly and not merely incidentally—a true interpersonal therapeutic relationship: «This pathic quality may well constitute the core meaning of the healing act of nursing care. If, as patient, I trust this hand then it has the power to reunite me pathically with my body, reminds me that I am one with my body, and thus makes it possible for me to heal, to strengthen, to become whole» (Van Manen 1999, p. 33).

The spiritual meaning of the caress, the fact of expressing tenderness or affection towards the other, constitutes the true essence of the caress, the fundament of every individual and fully human caress, as the phenomenology of the caress developed by José Gaos convincingly shows in a text of unsurpassable beauty and fine description. So much so, that even «in the case of sexual love and the caresses within it, these caresses are not an expression of sexual love, they are something non-sexual in sexual love» (Gaos 2010, p. 61). In the philosophy of the caress proposed by Gaos «[a] caress is intimacy between people as such» (Gaos 2010, p. 68), creating a space of mutual acceptance in which affection is transmitted. Thus, surprisingly, the personal pathic touch and private pathic touch share the same deep root, an important trait that Van Manen does not accentuate, possibly because he is more preoccupied with indicating as clearly as possible the boundaries between the professional sphere and the private sphere. Gaos attaches great spiritual significance to the caress, whether it be in the professional field, as with the relationship between nurse and patient, or in the personal domain of two people who love each other, a way of expressing the appropriate affection for a carnal being, the most subtle and immaterial form of material love.

## 4 Conclusions

Throughout this study devoted to the therapeutic hand, we have distinguished, in the first place, an instrumental or technical function, then a noetic or cognitive function and finally a pathic or affective function. Using this distinction, classical inspiration has been followed. Hippocrates, speaking of the excellence of the hands of the surgeon, referred to the «eurythmy of the hands» (*De habitu decendi*, L. IX, 236). Pedro Laín Entralgo, referring to any manual operation in the therapeutic field, breaks it down thus: «that excellence has within it three ingredients: a neuromuscular one, *eucinesia* or dexterity in the movement of the hands, an intellectual one, *eunoia*, the 'know-how' of the hand of the surgeon, as the executive organ of a scientifically–trained intelligence, and another which is moral, *eubulia*, the goodwill of one who lucidly seeks the good of another» (Laín 1964, pp. 343–344).

We have considered three different aspects of the hand, but in reality they act as a unity. The hands that intervene in the therapeutic act operate simultaneously under the three functions, although in certain medical or curative actions one or other aspect predominates. However, the pathic or affective function is, in our view, the very heart or essence of the therapeutic hand. In other words, it serves as the very foundation. In effect, the therapeutic hand, generally considered, is first and foremost a hand that expresses concern, care, attention, the will to heal, the determination for life and determination for the good of the other. The technical and diagnostic means it uses to achieve this end are overshadowed by the fundamental choice that is made—that is to seek the care, healing, restoration and well-being of another human life. The therapeutic hand is a hand that extends from a life to place itself at the service of another life and offer, as if it were a bestowal, a precious gift, something which can prove to be of benefit to it.

An action which is detrimental to the life of another, technically valid from the point of view of manual dexterity, can never be a therapeutic one. It could even be an act of torture, murder or sadism. Therefore, in the case of a therapeutic action, the pathic or affective hand must always be present, underlying or enveloping any other dimension. Furthermore, of the three functions indicated, the pathic or affective dimension is not just the one which is foremost, but the only truly irreplaceable dimension of them all. We can imagine in a not too distant date, which in some cases is already becoming a reality, the technical or cognitive dimension of the hand might be supplanted by technological devices. Now, no machine can substitute the warmth, immediacy and emotionality of a hand that cares and signals the beneficial presence of another human being. The placing of any technological buffer between them would imply, in this field, the imposition of a distance that would deny the very thing that one wishes to transmit.

Perhaps for this reason, an author like José Gaos considers the defining characteristic of the human hand to be the possibility it possesses for caressing: «I would say: it is not simply that the hand can caress but: it is the possibility to caress what makes or creates the hand» (Gaos 2010, p. 58). He is not alone in his thinking. Jacques Derrida, commenting on some passages in the work of Martin Heidegger in which he refers to the human hand, underlines the fact that Heidegger sees the essence of the hand in the act of giving: the defining aspect of the hand is giving and receiving, but not only of things, but also in giving itself and being received itself: the hand is given and receives itself in the other (Derrida 2010). Finally, an author like Emmanuel Lévinas, in analysing phenomenologically the essential elements of intersubjective relations, emphasises first the affective component. For him, the essence of a handshake or a caress is the trust, the bond and the peace transmitted, in the manner of a gift, from one human being to another (Lévinas 1978).

In short, the proteic versatility of the human hand makes it an admirable body organ for the innumerable uses and actions to which it lends itself. In the field of health care, despite a necessary reduction of its initial possibilities and an inevitable specialisation of its tasks, the hand maintains its characteristic multifunctionality. In fact, the therapeutic hand is, all at the same time, instrumental, cognitive and affective. Of all these features, the one that appears as the least technical or least specialised, that is the pathic hand, has, nevertheless, an underlying role with respect to the others. The pathic or affective dimension is the very basis of the therapeutic hand, since it is this that drives it to act for the good of the patient, and it is this dimension that in some way must direct and involve any operation of the therapeutic hand, whose ultimate aim is the care and healing of the human person.

# References

Aristóteles (2003). *Sobre el alma*. Madrid: Ed. Gredos.
Aristóteles (2014). *Ética a Nicómaco*. Madrid: Centro de Estudios Políticos y Constitucionales.
Arnould-Taylor, W. E. (1991). *The principles and practice of physical therapy*. Cheltenham: Stanley Tomas.
Barbotin, M. (1970). *L'humanité de l'home*. Paris: Aubier.
Bell, S. C. (1979). *The hand: Its mechanism and vital endowments as evincing design*. Cleveland: Pilgrim Press.

Billhult, A., Bergbom, I., & Stener-Victorin, E. (2007). Massage relieves nausea in women with breast cancer who are undergoing chemotherapy. *The Journal of Alternative and Complementary Medicine, 13*(1), 53–58.
Buytendijk, F. J. J. (1970). 'Some aspects of touch', *Journal of Phenomenological Psychology, 1* (1), pp. 99–124.
Cole, J. (2013). 'Capable of whatever man's ingenuity suggests': Agency, deafferentation, and the control of movement. In Z. Radman (Ed.), *The hand, an organ of the mind* (pp. 3–25). Massachusetts: MIT.
Cooper, K., Lindsay, A., Hancock, E., & Smith, F. W. (2013). The use of pMRI to validate the identification of palpated bony landmarks. *Manual Therapy, 18*(4), 289–293.
Day, M., & Napier, J. (1963). The functional significance of the deep head of flexor pollicis brevis in primates. *Folia Primatologica, 1*(2), 122–134.
de Hennezel, M. (1996). *La muerte íntima*. Barcelona: Plaza & Janés.
Derrida, J. (2010). *Heidegger et la question*. Paris: Flammarion.
Elkiss, M. L., & Jerome, J. A. (2012). Touch—More than a basic science. *The Journal of the American Osteopathic Association, 112*(8), 514–517.
Farmer, H., & Tsakiris, M. (2013). Touching hands: A neurocognitive review of intersubjective touch. In Z. Radman (Ed.), *The hand, an organ of the mind* (pp. 103–130). Massachusetts: MIT.
Fingleton, C., Dempsey, L., Smart, K., & Doody, C. (2014). Intraexaminer and interexaminer reliability of manual palpation and pressure algometry of the lower limb nerves in asymptomatic subjects. *Journal of Manipulative and Physiological Therapeutics, 37*(2), 97–104.
Gadamer, H. G. (1996). *The enigma of health*. Stanford, California: Stanford University Press.
Galeno (2010). *Del uso de las partes*. Madrid: Ed. Gredos.
Gallagher, S. (2013). The enactive hand. In Z. Radman (Ed.), *The hand, an organ of the mind* (pp. 209–225). Massachusetts: MIT.
Gaos, J. (2010). La caricia. In A. S. de Haro (Ed.), *Cuerpo vivido* (pp. 53–85). Madrid: Ed. Encuentro.
Gehlen, A. (1987). *El hombre*. Salamanca: Sígueme.
Husserl, E. (1952). *Ideen zu einer reinen phänomenologie und phänomenologischen philosophie. Zweites Buch. Phänomenologische Untersuchungen zur Konstitution*. The Hague: Martinus Nijhoff.
Jonas, H. (1954). The nobility of sight. *Philosophical and Phenomenological Research, 14*(4), 507–519.
Jonas, H. (1998). *Pensar sobre Dios y otros ensayos*. Barcelona: Herder.
Laín Entralgo, P. (1964). *La relación entre médico y paciente*. Madrid: Revista de Occidente.
Lévinas, E. (1978). *Hors sujet*. Montpellier: Fata Morgana.
Mattens, F. (2013). Perception and representation: Mind the hand! In Z. Radman (Ed.), *The hand, an organ of the mind* (pp. 159–184). Massachusetts: MIT.
McCorkle, R., & Jeannie, V. P. (2001). Enhancing caregiver outcomes in palliative care. *Cancer Control, 8*(1), 36–45).
Merleau-Ponty, M. (1945). *Phénoménologie de la perception*. Paris: Gallimard.
Merleau-Ponty, M. (1964). *Le visible et l'invisible*. Paris: Gallimard.
Minkowski, E. (1999). *Vers une cosmologie*. Paris: Ed. Payot.
Peig, C. (2007). *Génesis del concepto de trabajo en Santo Tomás. Su contexto histórico y doctrinal*. Excerpta e Dissertationibus in Sacra Theologia (Vol. L, n. 2). Pamplona: Facultad de Teología de la Universidad de Navarra.
Pera, C. (2003). *El cuerpo herido. Un diccionario filosófico de la cirugía*. Acantilado: Barcelona.
Polo, L. (1991). *Quién es el hombre*. Madrid: Rialp.
Radman, Z. (2013a). *The hand, an organ of the mind*. Massachusetts: MIT.
Radman, Z. (2013b). On displacement of agency: The mind handmade. In Z. Radman (Ed.), *The hand, an organ of the mind* (pp. 369–397). Massachusetts: MIT.

Roesch, E. B. (2013). A critical review of classical computational approaches to cognitive robotics: case study for theories of cognition. In Z. Radman (Ed.), *The hand, an organ of the mind* (pp. 401–419). Massachusetts: MIT.

Routasalo, P. (1999). Physical touch in nursing studies: a literature review. *Journal of Advanced Nursing, 30*(4), 843–850.

Routasalo, P., & Isola, A. (1996). The right to touch and be touched. *Nursing Ethics, 3*(2), 165–176.

Strauss, E. (1952). The upright posture. *Psychiatric Quarterly, 26*(1–4), 529–561.

Tocheri, M., Orr, C., Jacofsky, M., & Marzke, M. (2008). The evolutionary history of the hominin hand since the last common ancestor of Pan and Homo. *Journal of Anatomy, 212*(4), 544–562.

Uvnäs-Moberg, K., & Petersson, M. (2010). Role of oxytocin and oxytocin related effects in manual therapies. In H. H. King, W. Jänig, & M. M. Patterson (Eds.), *The science and clinical application of manual therapy*. Amsterdam: Elsevier.

Valéry, P. (1993). *Estudios filosóficos*. Madrid: Ed. Visor.

Van Manen, M. (1999). The pathic nature of inquiry and nursing. In I. Madjar, & J. A. Waltonton (Eds.), *Nursing and the experience of illness* (pp. 17–35). London, New York: Routledge.

# Cooking and Human Evolution

**Maria Pia Chirinos**

**Abstract** According to anthropology and paleontology, the first division of work was between man-the-hunter and woman-the-gatherer. Both activities are cultural human features (process of humanization) that started after the morphology of human body came to an end (process of hominization). At the same time, the study of human evolution has focused mainly on paleontological theses and has explained the different stages based on evidence such as fossils, tools, and meat diet. Recent studies raise doubts about these hypotheses and propose a new assessment: prior to hunting and gathering, cooking could have been the key activity that made our digestive system so different from other primates and also explains our social behavior better. This implies that the control of fire could have taken place earlier than posited by archeology. In addition to this statement, which is supported by biological evidence, other theses enable us to chart anew the way cooking influenced the ordinary lives of our ancestors and the way human beings cared for one another.

**Keywords** Cooking · Division of labor · Human evolution · Manual work · Social behaviour · Richard Wrangham

Considering manual work as human work is a controversial topic. As Richard Sennett has stated, technology in the form of automation displaces not only blue-collar workers but even white-collar workers as human labor (Sennett 2006). If that is the case for these types of work, cooking occupies an even worse position. But many years ago, Yves Simon in his book on work stated: "a good method requires that the search for the definition should start with the most unmistakable case. Accordingly, we shall first consider the case of the manual workers, rather than the lawyer, merchant or scholar" (1986, p. 4). It is not the aim of this study to debate the justice of this polemical issue. Its purpose is to define the proper place in human evolution of a type of manual work that has not been present in the debate

---

M.P. Chirinos (✉)
School of Humanities, Universidad de Piura, Piura, Peru
e-mail: mariapia.chirinos@udep.pe

until this century, and to argue that it plays a more central role than previously recognized: the work of cooking.

## 1 The Process of Hominization Versus the Process of Humanization

The literature about human evolution does not use the two terms "hominization" and "humanization." This does not, however, imply that their meaning is absent. Pierre Teilhard de Chardin introduced a reference to hominization in the 1950s (Teilhard de Chardin 1955). Some years later, authors like Choza (1988), Polo (1992) and Malo (1999) distinguished between hominization and humanization and, in doing so, offered a useful new terminology. Hominization refers to the morphological changes which took place among the chain of hominids and resulted in the human body with all its specific characteristics: bipedalism, the size of the cranium, the movement of the hands, etc.

The process of humanization, on the other hand, is specifically related to those forms of behavior which begin to differ from what is merely animal. Daily and bodily needs require fulfilling in ways that are no longer guided solely by instincts. Culture begins as an effect of this process, as a result of the emerging of new human dimensions such as intelligence, freedom, etc. As Polo explains, hominization has to do with our bodily reality, and humanization with our psychological–cultural dimension (1992).

As already mentioned, in other cultural areas, the two words are not frequent. Philosophers, scientists, anthropologists use the distinction without using the words. Frederick Engels, for example, stated that cooking and the domestication of animals, "directly became new means of emancipation for man" (1975 [1876], p. 9).

Charles Darwin took it for granted that morphological changes, which led to our present configuration, had taken place before man learned to control fire. According to him, how to use fire should probably be considered the greatest discovery ever made by man (Darwin 1871). For Darwin, then, first came hominization, or the finishing of the human body, and then humanization, which enabled human beings to behave differently.

Another great anthropologist and ethnologist, Claude Lévi-Strauss (1969), contributed similar theses that were revolutionary in his time. In his view, cooking is a strictly cultural factor and implies a leap forward with regards to nature. Edmund Leach, a disciple of Lévi-Strauss, described cooking as a symbolic way of showing that we are not beasts (Leach and Claude 1970).

In other words, all these authors defended the thesis that once morphological changes were completed (the process of hominization), the new *Homo* began to behave according to different parameters (the process of humanization). When nature has finished changing our body and culture begins to transform our world, hominization stops and humanization starts. In this connection, the first question I want to discuss in this paper is whether these two processes really function in that way, one after the other, or not.

This would be the first or strict sense of humanization. But nowadays we could give the term a wider meaning. The second meaning of "humanization" appears to be due to a serious "deficit" caused by a critical "surplus." Manual work like cleaning, cooking, etc., and others like attending customers or even teaching, once thought of as exclusive to human beings, can now be performed by sophisticated machines. The more technology advances, the less human work is needed, because it is replaced by high technology (Sennett 2006). Furthermore, the more technology facilitates communication, the less we can develop empathy for others (Sennett 2012). As a result, communities that are more developed in economic terms lack interpersonal relationships and appear to have a deficit in caring.

In this new situation, man and woman face the great challenge of recovering what was one of their first features, cooperation and interdependency, through daily manual tasks. We have to rediscover maybe the first result of the humanization process in its strict sense: the household as the environment which favors caring for others, based on a very human and ancient activity, manual work.

## 2 The Debate Around the Origin of the Division of Labor

Women's commitment to the household and men's to hunting—i.e., the "division of labor"—is a vexed question. As asserted by Lupo and Kiahtipes (Gurven and Hill 2009, p. 64), "understanding how, when, and why the sexual division of labor emerged is one of the most compelling issues in human evolution."

In the 1960s, Morris (2010 [1967]) emphasized that hunters preferred to take the meat to their families rather than eating it by themselves. The hunter focused his attention on the relationship between him and his female partner, and on his offspring's survival. As a result, human beings, among all primates, won the "evolution race". Washburn and Lancaster (1968) also contributed to this view when they wrote that "family organization may be attributed to the hunting way of life" (Washburn and Lancaster 1968, p. 295). That is, household life, for which women were especially responsible, appeared as the result of a deficit: the inability to go hunting, particularly if they were pregnant or had to care for their offspring. In other words, this task resulted from a condition of inferiority. And this division would be at the base of the nuclear family.

The feminist reaction against the man-the-hunter model appeared in the 1970s and 1980s. Abundant literature centered around defending the role of the woman-the-gatherer and proclaiming her equality with man-the-hunter (Dahlberg 1981; Zihlman 1989). The main idea was to demonstrate that hunting was not the principal or key task that caused the division of labor. Primitive societies needed both activities, and more attention needed to be paid to the influence of woman-the-gatherer. To cite but one example, the over-exploitation of the grounds where women dedicated themselves to gathering food was held to be the main reason for nomadic populations to move to new places (Kelly 1995).

The debate resulted in a lively exchange of views which did not lead to a solution, but enriched knowledge of women's activity in comparison to men. For example, from 179 hunter–gatherer societies examined by Carol Ember in a well-known study (1978), men alone hunted in 166 while men and women hunted together in 13. In none did women alone hunt, whereas women were the main gatherers in two-thirds of these societies. Among the most famous examples are the Agta women of the Philippines, who hunted with bows and arrows and killed the same prey as men (Estioko-Griffin and Griffin 1981). At the same time, according to these authors, there were many critical aspects that supported a sexual division of labor, without implying a gender bias. Among others, there were sex-differentiated comparative advantages in tasks, but also the need for a healthy diet provided by different foods whose "acquisition requires separate skills and additional time investment in learning and practice for increasing returns" (Gurven and Hill 2009, p. 57). These findings, as the authors explain, are broadly consistent with Gary Becker's argument concerning the family division of labor applied to hunter–gatherers (Becker 1973, 1974). In short, the dilemma between hunting and gathering is groundless: "specialization does not imply that men should hunt and women gather" (Gurven and Hill 2009, p. 57).

Taking this premise into account, the second issue I want to address is the question of the "monopoly" of these two activities, hunting and gathering, as the two main manifestations of the division of labor. In other words, is there a way out of this dilemma? Are these activities really the key actions that define primitive societies and their ordinary life?

According to a quite revolutionary study on fire and cooking, Richard Wrangham (2009) proposed a new approach to this problem: hunting is only possible if sufficient calories have previously been consumed. Only then can the hunter undertake the tiring and long-lasting activity of hunting. An increase of calories in the diet comes first from the consumption of meat, but secondly from the consumption of cooked food. Thus, cooking, more than hunting or gathering, is a key activity in human evolution.

Since Wrangham's book appeared, his thesis has been subjected to intense criticism and consequently to still more intense debate. As a result, not only has the scientific community developed a good number of studies, many of them guided by Wrangham and other researchers, but it has also undoubtedly strengthened and improved his thesis.

## 3  Brief Overview of Some Theses Regarding Human Evolution

Before referring to Wrangham's thesis in more detail, it is worth remembering other theses on human evolution. Until recently, paleontology seemed the most influential science in defining the milestones of the whole evolution issue. Based on the study of fossils, it has looked at human morphology, hominization, but also the early work of transforming stones and branches and, with it, the making of tools and weapons used in hunting. In other words, paleontology has also identified our rational behavior—when humanization begins. The point here is to make it clear what paleontology has defined as the first signs of humanization: they are not fossils but tools found next to fossils, which are hunting tools. This fact explains the very first hypothesis about man-the-hunter mentioned above.

What are the main milestones for paleontology? I will make a brief summary, which will necessarily include some that are already well-known. We can begin by examining a species previous to *Homo*, the *Ardipithecus ramidus* (Johanson and Shreeve 1989; Shreeve 2010). The existence of this species was deduced from some well-known footprints in ash, found by Leakey (1979) in Laetoli, Tanzania. The footprints were certainly identical to current human ones, undoubtedly from a biped. But the possibility of finding not footprints but remains from this species arose in Afar, the rich deposit in Eastern Africa. About 4.4 million years ago the *Ardipithecus ramidus* lived there as a quadruped in trees and as a biped on land. Their herbivorous diet is shown by small canines both in men and in women. *Ardipithecus*' erect and biped stance might be the oldest identifying link in the human phylogenetic tree (Johanson and Shreeve 1989).

A further milestone in human evolution is represented by Lucy, the most famous *Australopithecus*. Her biped walking reveals a key component in her body: the *foramen magnum* or orifice located in the skull, through which the spinal cord goes. Because of this, her head is positioned at the top of the spinal cord and her backbone forms a straight line with her legs. Among the primates sampled, humans exhibit the most anteriorly positioned *foramina magna*, which is an indicator of bipedal locomotion in fossil hominins (Russo and Kirk 2013). This verticality also requires modified musculature for the new position, which little by little will prevent Lucy from climbing trees and jumping from one tree to another. Besides facilitating the act of walking, bipedalism allows Lucy freer use of her upper limbs; claws, typical of quadrupeds, no longer need to grasp branches. Their use for defense or attack starts to decline. Everything is ready for the human hand to emerge, with its well-known features (Young 2003): the human thumb appears longer, the palm and fingers are shorter, the fingers lose their curvature (Susman and Creel 1979), and the ulnar position of the thumb is also very flexible (Pryce 1980). As Napier (1960) suggested, human hands are unique. They differ profoundly of those of chimpanzees.

The new biped form in Lucy bears a direct relationship to her diet. Many authors agree that the first hominids rapidly become carnivorous precisely because of their erect position (Walker 2007). Lucy was in fact omnivorous: she started to eat meat, and the increase in energy provided by the new diet brought about an increase of the skull capacity and also of the body size (among others, Aiello and Antón 2012): from a very small brain similar to those of apes (402 cm$^3$), the brain grew to a capacity of 699 cm$^3$, which has been called the "Brain Rubicon" (Montagu 1961). The erect position and locomotion demanded more calories, proteins and fats, which contributed to developing the brain size, one of the most "expensive" organs from a metabolic perspective (Aiello and Wheeler 1995).

The next paleontological milestone shows us *Homo habilis*, 2.6 million years ago (Johanson et al. 1987). The discovery was surprising not only because of its anatomy (hominization), but especially because of a new finding: for the first time, simple and naturalistic branches and stones appeared beside the fossils (beginning of humanization) (Leakey et al. 1964). *Homo habilis*, indeed, owes its name to his ability to elaborate primitive and intuitive weapons: the use of stones with sufficient cognitive exercise to kill large animals.

According to Aiello and Antón (2012), considerable attention has been paid to this link in the human evolution chain. Anatomically, the hand of *Homo habilis* resembles that of modern humans (Napier 1960, Ambrose 2001) and its brain, as we have already said, is significantly larger (600–800 cm$^3$) than that of australopithecines. Its left hemisphere has an impression of Broca's area which controls oro-facial fine motor control and language (Greenfield 1991). Finally, its teeth are relatively small for its body size (Pilbeam and Gould 1974). These three features suggest a relation between tool use, quality of diet, and intelligence.

The last milestone I will refer to is *Homo ergaster* or worker, better known as *Homo erectus*, because he was the first who left Africa and walked to distant territories. Both names are appropriate to describe this new species. First, because the most likely interpretation is that *Homo habilis*, although bipedal when terrestrial, still engaged in frequent arboreal behavior, while *Homo erectus* was a completely committed terrestrial biped (Ruff 2009). But the name *ergaster* is also applicable because of the presence of more sophisticated weapons and tools than those used by his predecessor, which enabled him to hunt bigger and more dangerous animals. According to paleontology, this feature added more meat to his diet (Aiello and Wheeler 1995, Aiello and Key 2002). Several authors have related this quantitative increase to the notable growth in the size of his brain, reaching 1240 cm$^3$. Ralph Holloway (1983) also points out his social behavior, basically more cooperative than aggressive, and therefore with the best conditions to start the division of labor.

Before ending this section, we can focus on the brain growth. To what extent is its size relevant? Increasing brain size, according to Schoenemann (2006), leads to increasingly complex processing within areas, greater degrees of autonomy and possible interactions too. But there is still more: all these features not only reveal a direct relation to conceptual complexity, but also suggest that brain evolution may be explained by the coevolution of language (Deacon 1997). And this is closely related to the socially interactive nature of humans, as well as the general association between degrees of sociality (Schoenemann 2006). Brain growth is therefore a relevant element. The point here is to determine the main cause of this process.

Paleontology points to a carnivorous diet as being the chief influence in brain growth (Bunn 2007), which indirectly implies that hunting, as a means to acquiring meat, is a key activity in human evolution.

## 4 Does This Explanation Offer Sufficient Support for the Thesis of the Evolution of the Different Hominids to *Homo Sapiens*?

While there is a general agreement on the causes that lead Lucy toward *Homo habilis*, i.e., her carnivorous diet, the supposition that an increase in the amount of meat (Milton 1999) or in its variety (Ungar et al. 2006) is at the base of the next step —*Homo habilis* to *Homo erectus* or *Homo ergaster*—does not seem so convincing. Research on carnivorous diet at the different stages of evolution has mostly ignored a factor which Wrangham (2009) stresses for the first time: not the type of food, but the fact that the food is cooked. This hypothesis about *Homo habilis* implies a bold assertion: that the dominion over fire goes back to the low Paleolithic stage, between 2.6 and 1.9 million years, which reveals the non validity of archeology's golden rule: "as a discipline [archeology] still has difficulties with the conundrum of 'absence of evidence is not evidence of absence'" (Gowlett and Wrangham 2013, p. 10). The starting point of the dominion over fire by men or women would be difficult to determine because of the very nature of fire, which destroys all evidence.

But the hypothesis of an earlier beginning for the use of fire is of high importance, because as many authors have written in different ways, but usually referring to fully evolved human beings, "the capture of fire by the genus *Homo* changed forever the history of the planet. Nothing else so empowered hominids, and no other human technology has influenced the planet for so long and so pervasively" (Pyne 1997, p. 12).

In a later study, Gowlett and Wrangham, in view of the difficulty of reconciling their proposals about the beginning of fire-use with archeological principles, make a notable effort to suggest meeting points between the different sciences, starting from some soundly based archeological facts. First, the distinction between natural fire and human fire, and the admission that the former came first and occurred in areas frequented by humans. Second, hearths are the principal archeological indicator of fire-use, but not the most widespread. Third, the most powerful evidence in discriminating natural and human-controlled fire is single artifacts or sets of artifacts showing very selective heating of humanly made material (Gowlett and Wrangham 2013, p. 10).

This last thesis represents a natural bridge to our issue: the place of cooking in evolution. References to cooking and human evolution, indeed, have always been scarce. In part, as already mentioned, because paleontology and archeology have led the research on evolution since the beginning (Aiello and Wheeler 1995, Bunn 2007). Nonetheless some writers have suggested a relation between cooked foods and the morphology of dentition and the intestine, reducing tooth size and gut size (e.g., Susman 1987, Aiello and Wheeler 1995). Additionally, none of the writers who have studied the importance of hunting or gathering has paid special attention to cooking. But it is still possible to take into consideration a silent and bodily testimony that can show the importance of this work: our peculiar digestive system. If this feature reveals any relation to cooking and to the dominion over fire, or human-controlled fire, then another science can offer answers to important evolutionary questions: biology (Wrangham 2009).

To summarize it briefly: if we compare our digestive system with that of chimpanzees, our nearest living relative, we discover that it presents unique features, from the sense of taste that helps us decide what to eat and influences how efficiently we digest these foods (Breslin 2013), to the morphology of each of its components. The facts that the shape of the human lips is considerably finer; the jaws weaker and of smaller size; and the teeth and molars are the smallest in terms of the body of all primates, are difficult to explain if our ancestors continued to eat raw meat (Ungar 2012). Moreover, the human stomach is a third of the size of the stomach of any other primate of our size, and the colon is 60% smaller than the expected size compared to other primates (Wrangham 2009). These are non-accidental numbers. Even before Wrangham, Katherine Milton admitted that "there is one striking difference between the anatomy of humans and apes. This difference is in the size relationship of different sections of the gut" (Milton 1999, 12). Applying fire to raw food helps with a task which was previously done by the body alone: softening raw food, facilitating digestion, etc. These costs of the digestive system would have diminished precisely due to cooked food. And therefore, an important cause for the reduction of our digestive system could be found in the fact that it no longer had to spend such a long time on the task of digestion.

And now we can go a step further: was cooked meat enough? What was the diet of the first hominids like? As Wrangham (2009) explains, meat composed up to 40% of the hunters' diet, a proportion much higher than the food of other carnivorous primates. This means that the hypothesis of "meat only" is not plausible for diverse reasons: among others, it does not explain the size of our teeth and, above all, if ingested in large quantities, meat contains a protein that is harmful to our health. Therefore, meat can only be part of the solution. We have to add two more elements: first, human beings, especially hunter-gatherers in arid areas, are the only ones whose diet adapts to a great proportion of starch and therefore many of them fed from a great number of tubers, grain, corn, etc. (Perry et al. 2007). Secondly, the variety in our foods is not random, but specialized: it demands organization and order; it shows improvement, transmission, and learning. Cooking becomes a sort of practical science; a proper human knowledge based on skills and also on theoretical truths that today can be called gastronomy (Chirinos 2006).

Another consequence of this thesis (Wrangham 2009) is that if a man dedicates most of the day or even several working days together, to hunting, he can only easily satisfy his hunger if his food is cooked and implicitly if he is the cook when he is hunting. This is because if the food he assimilates were all raw, he would have a great problem: like other primates, he would use approximately 5 h of every day to ingest and chew raw food with enough calories to keep him going. That would be almost 50% of the working day, which lasts approximately 12 h. Therefore, if the diet were raw, hunting would simply become non-viable. By contrast, the beginning of cooking as a daily activity transforms food into more energy, with more calories, being softer to digest and less astringent. The length of time devoted to eating is drastically reduced to 2 h per day. For all these reasons, Wrangham (2009) states, what makes man or woman unique is precisely his ability to process foods, in general, and to cook them, in particular.

According to Wrangham (2009), this new evolutionary opportunity would take place again in the next step, from *Homo erectus* to *Homo sapiens*, also thanks to the improvement in the cooking techniques such as the use of stones for primitive ovens, which would have increased the supply of edible foods and of new habitable areas (Fernández-Armesto 2002).

This all allows one to hypothesize that cooking, even when not representing a natural activity but a cultural one, was a decisive factor in making us physically human: the morphology of the digestive system and the consequences derived from it, influenced the growth of the brain and the consolidation of the bipedal position. The process of hominization, therefore, took place at the same time as the process of humanization.

## 5 Control of Fire, Food, and Some Habits Characteristic of the Human Being

To test the hypothesis of an earlier use of fire by humans, one would have to prove significant cognitive development, more than was required by the first stone technologies (Wrangham 2009). Indeed, cooking is not a simple activity. It implies complex behavior and multiple cognitive abilities. The simple fact that cooking involves the transformation of raw food into more desirable cooked food demands causal reasoning and future-oriented cognition (Warneken and Rosati 2015). In their article, Warneken and Rosati try to prove that, after simple, undemanding training, chimpanzees and humans share several of the essential psychological capacities needed to cook food: *inter alia*, they preferred cooked food over raw food, even though they need more time to acquire it; they show self-control by giving up food to transform it; after minimal experience, they exhibit a practical understanding of the basic transformation effected by cooking; and they generalize their knowledge about the technique to new types of food that they had never seen cooked. Chimpanzees also selectively attempt to place only edible items in the cooking device and finally, they transport their food to the cooking device and even save their food in anticipation of future cooking opportunities not immediately available.

The question is why is this important if they do not cook? Or even, why do they not cook? The answers are first, the importance of these results is that they suggest that, if chimpanzees acquire those capacities after minimal experience, this means that "early hominins may have also been able to detect and use existing opportunities in their environment to cook foods" (Warneken and Rosati 2015, p. 8). And the reason why chimpanzees do not cook is both the most obvious one: because they do not control fire; and, less obviously, because human cooking is generally social in nature. However, they add, "the social nature of cooking creates significant opportunities for theft, so increases in social tolerance may have been necessary for cooking to evolve" (Warneken and Rosati, 8; see also Wrangham et al. 1999).

At this point, archeology and biology can unite their scientific findings with the social hypothesis. A relevant proposal is related to some types of behavior which have been highlighted more clearly in recent years: cooperative upbringing in its wide sense, that is, the model according to which members other than parents (alloparents) within the social group help in the direct upbringing of children or help other mothers to raise them (Kramer and Otárola-Castillo 2015); and what is known as cooperative breeding (Burkart et al. 2009), which is not exclusive to humans, being also present in the callitrichids, a primate family in Central and South America. "The crucial feature here," as Burkart, Hrdy, and Van Schaik explain, "is the availability in the group of a number of reliable helpers" (Burkart et al. 2009, 178–179) that can currently be collocated with the emergence of the *Homo erectus* (Hrdy 2005; Burkart and Van Schaik 2010) and is simultaneous to a developed cognitive system that presents two characteristics: first, cognitive skills become available for deployment in cooperative contexts; and second, and more interestingly, the developed cognitive system amplifies opportunities for social learning (Burkart et al. 2009).

Upbringing and breeding are related to different actions such as caring and feeding, and not related to others such as fighting or leaving the offspring once they grow. Unlike animals, that fight for prey and in some cases look for food for their young, adult humans distribute food not only among their partner and offspring but also among other adults: as humans they recognize the need to care about others because of our dependent nature (MacIntyre 1999). The fact that *Homo* habitually has a sufficiently large quantity of food has been seen by many authors as the cause of a new human behavior which would later become characteristic: sharing it (Pontzer 2012, Isaac 1978). Indeed, "without sharing, large game hunting may never have been viable for our hominin ancestors" (Gurven and Hill 2009, p. 53).

Exchanges among families of all societies are an undoubted and constant fact (Isaac 1978). Each family represents a micro-economy which revolves around distribution, preparation, consumption, and conservation of food as a vital source. But there is something more: man-the-hunter does not hunt by himself but in a group, and the rule is that the group "decides" who keeps the day's prey, without it necessarily belonging to the skilled hunter who managed to kill it. There is therefore no strict relationship between killing and possessing, but rather between killing and distributing (Hawkes 2001, Wrangham 2009). Man-the-hunter has neither the right nor the obligation to take food home—a situation which can only be understood in light of the other side of the coin: the woman who cooks and organizes the domestic environment knows how to make cooked food; she administers and keeps it, and also knows that the family's food might depend mainly on the gathering she does each morning. "From ethnographic reports it seems that this domestic service is often the most important contribution a wife makes to their partnership" (Wrangham 2009, p. 167).

In effect, this other side of the coin goes hand in hand with a primitive right to property. Surprisingly, the group of women who go out each morning to gather fruits and plants exhibit behavior that is different from that of hunters. In primitive societies, as written many years ago (Schmidt 1935), the notion of property existed for all the "movable" objects resulting from individual activity or acquired as a gift. The rights of property of the children were respected by adults, as those of women were respected by men. "Thus women may freely dispose of the vegetables, fruits, etc., they have collected" (Schmidt 1935, p. 246). What is more, social peace in a field inhabited by different families depends on this first right to property: the woman is the owner of food and this right protects her against the demands of others. The fact that, in primitive cultures, the request to share food was symbolically associated to marriage, explains this behavior. Because of this, if a single woman offers food to another man, also single, or he requests food, this act in fact means and gives place to marriage. Any alteration of this basic norm disrupts social peace: a man who requests food from a married woman (who is not his wife) or a married woman who offers food to a man (who is not her husband) is in danger of being accused of adultery (Turnbull 1965, Collier and Rosaldo 1981, Wrangham 2009).

It is clear that only among human beings do we find this behavior, whose foundation is the woman's right of property regarding food and her dedication to cooking. Furthermore, we could state that the origin of the house, that is of the home and therefore of marriage and the family, cannot be fully understood without this particular behavior. Many have situated its origin in sexual intercourse, but this proposal could be amplified. For marriage to take place it is necessary to grasp a notion that we often take for granted but which is crucial: only humans are capable of a reciprocal, unique and permanent relationship, which implies bodily giving and the caring for the other. This perception, absent in the animal world, starts with a daily reality such as food, and not with any type of food, but with that which is the result of different types of manual work: cooking (including control of fire) and, at a second level, hunting and gathering.

But it should be added that, according to these presuppositions, it is not that sex does not count, but it is not enough: what makes it possible for the family to have appeared only among hominids is the awareness of notes which are absent in animals. Sex is not a human invention. Cooked food, caring for others or the control of fire are. These in part imply manual, free, and rational work capable of creating culture, as well as remoteness from instinctive behaviors (Chirinos 2006). Besides, contrary to the case of animals, the marriage pact appears in a very specific context, and traditions around food and the house support its emergence. Not in vain has ethnography always corroborated that, in most primitive societies, a married woman enjoys a high status of considerable autonomy (Wrangham 2009). Furthermore, "the rule that domestic cooking is women's work is astonishingly consistent" (Wrangham 2009, p. 151) and cooking appears as a social activity, which calls for defined relationships and which supports and reinforces social norms.

By this work of domestic cooking, women developed the first economic activity. Gathering, storing, and preparing foods, through the control of fire, constitute an anthropological universal which leads us to recognize a still more important one: the fact of our dependence on others, due to having bodily needs that cannot be supplied by ourselves, both during our childhood and in our old age, especially when we are sick (MacIntyre 1999). The house in its human dimension becomes the place where family originates and develops as a point of reference to which its members come back. And within them, in the first place, man-the-hunter who comes back, with or without hunted prey, looking for food to regain his strength. However, politically incorrect, we have to admit that woman-the-cook is at the basis of the division of work and of family social life. Without her, man-the-hunter would not have been able to carry out his activity fully, and therefore, probably, the division of work would not have appeared.

# 6 Conclusion

The division of work is still an open question, and debate has reappeared in the academic field since Wrangham (2009) proposed his theses about cooking and the use of fire. Some consequences of this approach are:

– A new understanding of both processes: of hominization and of humanization. Against some relevant studies and authors, it seems that they are not two different moments. The process of humanization does not have to wait for the complete development of our body in order to start. Changes in our morphology seem to be effects of changes in our conduct and vice versa. Biology is the science that can now be used to explain this.
– Cooking manifests the process of humanization because it implies control of fire and its use for many purposes, not present in other animals. It has also had important influences in our process of hominization. Although theoretically we could eat raw food and survive, recent studies state that having started to ingest cooked food has been determinant in human morphological changes such as our digestive system.
– The dedication of women to cooking is a surprisingly consistent activity in primitive cultures. Cooking (with prior control and use of fire) seems to be the human universal which could explain the division of labor.
– To cook implies not only many cognitive conditions, more difficult to develop than the making of tools, but is also related to cooperative upbringing and cooperative breeding, which imply social attitudes. Both cognitive conditions and cooperative upbringing and breeding could be found in some families of

primates, which explain why humans could learn so rapidly to use fire for cooking and to share cooked food with others as a caring manifestation.
– These coincidences lead us to understand collaborative work in a broad sense. Although it is true that this way of working is also present in hunting, in the case of feeding and cooking, it reveals a different aspect: the ability to care for others, as a consequence of our dependence from birth onwards.

Obviously, stating this would imply that woman in the kitchen—woman-the-cook—occupies a place that anthropology has accorded to man-the-hunter in the division of work and gender studies: woman-the-gatherer. The thesis we propose in this paper, following Wrangham's research, maybe seems bold, especially for a sector of feminism, but it intends to contribute to an intense debate based on scientific evidence. It recalls Yves Simon's words: "a good method requires that the search for the definition should start with the most unmistakable case" (1986, p. 4).

## References

Aiello, L. C., & Antón, S. C. (2012). Human biology and the origins of Homo: An introduction to supplement 6. *Current Anthropology, 53*(suppl. 6), S269–S277.
Aiello, L. C., & Key, C. (2002). Energetic consequences of being a Homo erectus female. *American Journal of Human Biology, 14*(5), 551–565.
Aiello, L. C., & Wheeler, P. (1995). The expensive-tissue hypothesis: The brain and the digestive system in human and primate evolution. *Current Anthropology, 36*(2), 199–221.
Ambrose, S. H. (2001). Paleolithic technology and human evolution. *Science, 291*(5509), 1748–1753.
Becker, G. S. (1973). A theory of marriage: Part I. *Journal of Political Economy, 81*(4), 813–846.
Becker, G. S. (1974). A theory of marriage: Part II. *Journal of Political Economy, 82*(2, Part 2), S11–S26.
Breslin, P. A. (2013). An evolutionary perspective on food and human taste. *Current Biology, 23*(9), R409–R418.
Bunn, H. T. (2007). Meat made us human. In P. S. Ungar (Ed.), *Evolution of the human diet: The known, the unknown, and the unknowable* (pp. 191–211). Oxford and New York: Oxford University Press.
Burkart, J. M., & Van Schaik, C. P. (2010). Cognitive consequences of cooperative breeding in primates? *Animal Cognition, 13*(1), 1–19.
Burkart, J. M., Hrdy, S. B., & Van Schaik, C. P. (2009). Cooperative breeding and human cognitive evolution. *Evolutionary Anthropology: Issues, News, and Reviews, 18*(5), 175–186.
Chirinos, M. P. (2006). *Claves para una antropología del trabajo*. Pamplona: Eunsa.
Choza, J. (1988). *Manual de antropología filosófica*. Madrid: Rialp.
Collier, J. F., & Rosaldo, M. Z. (1981). Politics and gender in simple societies. In S. Ortner & H. Whitehead (Eds.), *Sexual meanings*. New York: Cambridge University Press.
Dahlberg, F. (1981). *Woman the gatherer*. New Haven: Yale University Press.
Darwin, C. (1871). *The descent of man and selection in relation to sex*. London: John Murray.
Deacon, T. (1997). *The symbolic species: The co-evolution of language and the human brain*. London: Allen Lane.
Ember, C. R. (1978). Myths about hunter-gatherers. *Ethnology, 17*(4), 439–448.

Engels, F. (1975). *The part played by labor in the transition from ape to man*. Peking, China: Foreign Languages Press (Original work published 1876).
Estioko-Griffin, A., & Griffin, P. B. (1981). Woman the hunter: The agta. In F. Dahlberg (Ed.), *Woman the gatherer* (pp. 121–151). New Haven, CT: Yale University Press.
Fernández-Armesto, F. (2002). *Near a thousand tables: A History of food*. New York: Simon and Schuster.
Gowlett, J. A., & Wrangham, R. W. (2013). Earliest fire in Africa: Towards the convergence of archaeological evidence and the cooking hypothesis. *Azania: Archaeological Research in Africa, 48*(1), 5–30.
Greenfield, P. M. (1991). Language, tools and brain: The ontogeny and phylogeny of hierarchically organized sequential behavior. *Behavioral and Brain Sciences, 14*(4), 531–551.
Gurven, M., & Hill, K. (2009). Why do men hunt? A reevaluation of "man the hunter" and the sexual division of labor. *Current Anthropology, 50*(1), 51–74.
Hawkes, K. (2001). Is meat the hunter's property? Big game, ownership, and explanations of hunting and sharing. In C. B. Stanford & H. T. Bunn (Eds.), *Meat-eating and human evolution* (pp. 219–236). Oxford: Oxford University Press.
Holloway, R. L. (1983). Human paleontological evidence relevant to language behavior. *Human Neurobiology, 2*(3), 105–114.
Hrdy, S. B. (2005). Cooperative breeders with an ace in the hole. In E. Voland, A. Chasiotis, & W. Schiefenhövel (Eds.), *Grandmotherhood: The evolutionary significance of the second half of female life* (pp. 295–317). New Brunswick: Rutgers University Press.
Isaac, G. L. (1978). The Harvey lecture series, 1977–1978. Food sharing and human evolution: archaeological evidence from the Plio-Pleistocene of east Africa. *Journal of Anthropological Research, 34*(3), 311–325.
Johanson, D. C., & Shreeve, J. (1989). *Lucy's child: The discovery of a human ancestor*. New York: William Morrow and Co.
Johanson, D. C., Masao, F. T., Eck, G. G., White, T. D., Walter, R. C., Kimbel, W. H., et al. (1987). New partial skeleton of Homo habilis from Olduvai Gorge. *Tanzania. Nature, 327* (6119), 205–209.
Kelly, R. L. (1995). *The lifeways of hunter-gatherers: The foraging spectrum*. Washington: Smithsonian Institution Press.
Kramer, K. L., & Otárola-Castillo, E. (2015). When mothers need others: The impact of hominin life history evolution on cooperative breeding. *Journal of Human Evolution, 84*, 16–24.
Leach, E. S., & Claude, E. L. (1970). *Lévi-Strauss, antropólogo y filósofo*. Barcelona: Anagrama.
Leakey, M. D. (1979). Footprints in the ashes of time. *National Geographic, 155*(4), 446–459.
Leakey, L. S., Tobias, P. V., & Napier, J. R. (1964). A new species of the genus Homo from Olduvai Gorge. *Nature, 202*, 7–9.
Lévi-Strauss, C. (1969). *The raw and the cooked: Introduction to a science of mythology, vol. 1.* (trans: John and Doreen Weightman). New York: Harper and Row.
MacIntyre, A. (1999). *Dependent rational animals: Why human beings need the virtues*. Chicago: Open Court Publishing.
Malo, A. (1999). *Antropologia dell'affetività*. Rome: Armando Editore.
Milton, K. (1999). A hypothesis to explain the role of meat-eating in human evolution. *Evolutionary Anthropology Issues News and Reviews, 8*(1), 11–21.
Montagu, A. (1961). The "cerebral rubicon": Brain size and the achievement of hominid status. *American Anthropologist, 63*(2), 377–378.
Morris, D. (2010) [1967]. *The naked ape: A zoologist's study of the human animal*. New York: Random House.
Napier, J. R. (1960). Studies of the hands of living primates. *Proceedings of the Zoological Society of London, 134*(4), 647–657.
Perry, G. H., Dominy, N. J., Claw, K. G., Lee, A. S., Fiegler, H., Redon, R., et al. (2007). Diet and the evolution of human amylase gene copy number variation. *Nature Genetics, 39*(10), 1256–1260.

Pilbeam, D., & Gould, S. J. (1974). Size and scaling in human evolution. *Science, 186*(4167), 892–901.
Polo, L. (1992). Sobre el origen del hombre. In F. Fernandez (Ed.), *Estudios sobre la encíclica Centesimus Annus* (pp. 110–121). Madrid: AEDOS.
Pontzer, H. (2012). Ecological energetics in early Homo. *Current Anthropology, 53*(S6), S346–S358.
Pryce, J. C. (1980). The wrist position between neutral and ulnar deviation that facilitates the maximum power grip strength. *Journal of Biomechanics, 13*(6), 505–507, 509–511.
Pyne, S. J. (1997). *World fire: The culture of fire on earth*. Seattle: University of Washington Press.
Ruff, C. (2009). Relative limb strength and locomotion in Homo habilis. *American Journal of Physical Anthropology, 138*(1), 90–100.
Russo, G. A., & Kirk, E. C. (2013). Foramen magnum position in bipedal mammals. *Journal of Human Evolution, 65*(5), 656–670.
Schmidt, W. (1935). The position of women with regard to property in primitive society. *American Anthropologist, 37*(2), 244–256.
Schoenemann, P. T. (2006). Evolution of the size and functional areas of the human brain. *Annual Review of Anthropology, 35,* 379–406.
Sennett, R. (2006). *The culture of the new capitalism*. New Haven and London: Yale University Press.
Sennett, R. (2012). *Together. The rituals, pleasures and politics of cooperation*. New Haven and London: Yale University Press.
Shreeve, J. (2010). Sur la route de l'evolution. *National Geographic*, julliet, 3–28.
Simon, Y. R. (1986). *Work, society, and culture*. New York: Fordham University Press.
Susman, R. L. (1987). Pygmy chimpanzees and common chimpanzees: Models for the behavioral ecology of the earliest hominids. In W. G. Kinzey (Ed.), *The evolution of human behavior: primate models* (pp. 72–86). Albany: State University of New York Press.
Susman, R. L., & Creel, N. (1979). Functional and morphological affinities of the subadult hand (OH 7) from Olduvai Gorge. *American Journal of Physical Anthropology, 51*(3), 311–331.
Teilhard De Chardin, P. (1955). *The phenomenon of man*. Brighton and Portland: Sussex Academic Press.
Turnbull, C. M. (1965). *Wayward servants: The two worlds of the African pygmies*. New York: Natural History Press.
Ungar, P. S. (2012). Dental evidence for the reconstruction of diet in African early Homo. *Current Anthropology, 53*(suppl. 6), S318–S329.
Ungar, P. S., Grine, F. E., & Teaford, M. F. (2006). Diet in early Homo: A review of the evidence and a new model of adaptive versatility. *Annual Review of Anthropology, 35,* 209–228.
Walker, A. (2007). Early hominid diets: Overview and historical perspectives. In P. S. Ungar (Ed.), *Evolution of the human diet: The known, the unknown, and the unknowable* (pp. 3–10). Oxford and New York: Oxford University Press.
Warneken, F., & Rosati, A. G. (2015). Cognitive capacities for cooking in chimpanzees. *Proceedings of the Royal Society B, 282,* 20150229.
Washburn, S. L., & Lancaster, C. (1968). The evolution of hunting. In R. B. Lee & I. DeVore (Eds.), *Man the Hunter* (pp. 293–303). Chicago: Aldine.
Wrangham, R. (2009). *Catching fire: How cooking made us human*. New York: Basic Books.
Wrangham, R. W., Jones, J. H., Laden, G., Pilbeam, D., & Conklin-Brittain, N. (1999). The raw and the stolen: cooking and the ecology of human origins 1. *Current Anthropology, 40*(5), 567–594.
Young, R. W. (2003). Evolution of the human hand: The role of throwing and clubbing. *Journal of Anatomy, 202*(1), 165–174.
Zihlman, A. (1989). Woman the gatherer: The role of women in early hominid evolution. In S. Morgan (Ed.), *Gender and anthropology: Critical reviews for teaching and research* (pp. 23–43). Arlington VA: American Anthropological Association.

# Essential to Art

## Hand, Touch and the Aesthetic Experience

Sixto J. Castro

**Abstract** In this article, I show how, from the very beginning, Aesthetics has opted for an intellectual consideration of the work of art, which allegedly corresponds to a pure inner experience. I explore how the external senses are minimized respect to the ideal creation and reception, with the sole exception of the "intellectual" ones—sight and hearing—being linked to the aesthetic experience. Among the "lower" senses—smell, taste and touch—the latter has been undervalued, and the role of human tactility has been considered ancillary. This has been important for the configuration of the modern concept of art. In exploring the relationship between aesthetic experience and the sense of touch, I demonstrate that prohibiting the appreciator to touch the artwork under his or her consideration is to perpetuate the same paradigm that considers the work of art an essentially intellectual reality. The review of these elements makes it necessary to rethink our current aesthetic categories as well as the concept of art itself.

**Keywords** Hand · Art (concept of) · Senses · Phenomenology · Aesthetics · Touch

## 1 The Constitution of the Modern System of the Arts

What today is often called "the modern system of the arts" is a complex of institutions, ideas, practices and theories that coincide with the rise of the modern philosophies of subjectivity. These philosophies hold that the subject is conceived primarily as a knowing being, and "true" knowledge is thought to be built on the "objective" model of science. In this context, the concept of art—traditionally considered as a series of rule-governed practices, as it was understood under both the Greek concept of "téchne" and its corresponding Latin "ars"—undergoes a redefinition within this new structure.

There are several reasons for this new conception of art. On the one hand, in the seventeenth century, the sciences were recognized as specific fields of knowledge

S.J. Castro (✉)
Departamento de Filosofía, Universidad de Valladolid, Valladolid, Spain
e-mail: sixto@fyl.uva.es

and practice, with their own methods and interests. The sciences brought experimental and mathematical methods together, and in so doing, they made themselves distinct from the liberal arts (Shiner 2001, p. 80 ff.). In addition to this separation from sciences, in order for the modern category of art—and subsequently the idea of Art (with a capital A)—to be defined, there must necessarily have been a reorganization of the qualities that until then had been part of the generic space of the arts understood as "téchnai". In the old system, it was supposed that in the artisan/artist there was a mixture of genius and rule, inspiration and work, innovation and imitation, freedom and service, etc. These pairs were separated, one from the other, in the late eighteenth century: poetic attributes (imagination, inspiration, freedom and genius) were assigned to the artist and the "mechanic" ones (skill, rules, imitation and service) to the artisan (Shiner 2001, p. 111 ff.). In this new system of the arts, only the artist—not the artisan—is credited with all abilities having to do with freedom. The artist is free from the imitation of the original models (originality), free from the dictates of reason and rule (inspiration),[1] free from the restrictions of fantasy (imagination) and free from the obligation to imitate nature (creation). Thus the ideology of genius, which will dominate the nineteenth century, is made possible, as well as the idea of "Art", that no longer designates the category of Fine Arts (poetry, painting...), but instead defines an autonomous realm of works and interpretations, values and institutions. Art would now refer to a kind of metaphysical space, and metaphysics is not brought about by the work of mere hands. Genius has no need of hands, and so Art does not need them either. Skilled hands will instead be the patrimony of craftsmen. Art is increasingly considered a fundamentally intellectual activity; even—as Hegel would have it—a rudimentary

---

[1] The dialectics between freedom and rule is present in the difficult and confusing passages that Kant devotes to the question of genius in the *Critique of Judgment*. Therein the oscillation between originality and regularity is constant. Kant claims that "genius is the innate mental predisposition *(ingenium)* through which nature gives the rule to art". Genius is the talent to produce what cannot be made by following rules. These "must be abstracted from what the artist has done" (Kant 1987, p. 177). The creative process is ineffable even for the artist. But Kant did not feel too comfortable with this "anomic" reality, and so, some paragraphs later, he tries to control it by subordinating the original power of genius to taste, submitting the freedom of imagination to the law of understanding. In order to stress the regulated character of artistic creation, Kant speaks against those who identify genius with "renouncing all rules of academic constraint, believing that they will cut a better figure on the back of an ill-tempered than of a training-horse. Genius –he claims– can only provide rich material for products of fine art; processing this material and giving it form requires a talent that is academically trained, so that it may be used in a way that can stand the test of the power of judgment" (*Ibid.*: 178). Thus, apart from the rule given by nature, there must be some other rules in art: "there is no fine art that does not have as its essential condition something mechanical, which can be encompassed by rules and complied with, and hence has an element of academic correctness" (*Ibid.*). But again, in the next paragraphs, Kant will insist on "imagination in its freedom from any instruction by rules" (*Ibid.*: 188) as a property of genius, and, at the same time, will claim that "when we judge art as fine art, is taste, at least as an indispensable condition (conditio sine qua non). In order [for a work] to be beautiful, it is not strictly necessary that it be rich and original in ideas, but it is necessary that the imagination in its freedom be commensurate with the lawfulness of the understanding. For if the imagination is left in lawless freedom, all its riches (in ideas) produce nothing but nonsense" (*Ibid.*). Kant seems to be afraid of his own anomic creation.

form of philosophy. That is why disregarding hands is essential to the formation of the modern concept of art.

This understanding can be seen, e.g. in the process of establishing the specific vocabulary of the discipline. In the field of nascent German Aesthetics in the eighteenth century, both the term *Gefühl*—which came to designate the realm of feelings and emotions—and the term *Empfindung*—which later described physiological sensation and feeling—referred primarily to the perceptual immediacy of a representation, i.e. to the sensuous aspect. Baumgarten—to whom we owe the name of our discipline (Aesthetics)—proposed to translate "tactus" by *Gefühl* (Baumgarten 2011: §536). Indeed, he quite indifferently used both *Tastsinn* and *Gefühl* instead of "tactus". Baumgarten's idea was to present a new discipline, Aesthetics, as a science of sensory knowledge (*cognitio sensitiva*), based on the alignment Wolff had established between *empfinden* (feeling) and *erkennen* (knowing). But Kant, in his *Critic of the Judgement* (§ 3), will set a radical separation between *Empfindung*—sensation, an objective representation (*Vorstellung*) of sense, equivalent to perception (*Wahrnehmung*)—and *Gefühl*—a representation referred solely to the subject and not available for any cognition. The example Kant gives is this: the green colour of a meadow belongs to objective sensation (*objektive Empfindung*), that is, to the perception of an object of sense, but its agreeableness belongs to subjective sensation (*subjektive Empfindung*) by which no object is represented, that is, to the feeling (*Gefühl*), through which the object is regarded as an object of our liking (which is not a cognition of it) (Kant 1987, p. 48) (Dubost 2014). From now on, the aesthetic judgment will appeal primarily to the subjective *Gefühl* and not to the objective sensation.

The consideration of art as a primarily intellectual, rather than manual, activity has a long history, even before the modern system of the arts. It was already present in that philosophical tradition which has probably been most decisive for the self-understanding of the arts in the Western world: neo-Platonism. In this tradition, the "idea" has absolute predominance (Panofsky 1968). Because of the close relationship between idea (*idéa*) and sight (*eidos*) in Platonic and neo-Platonic philosophy, the eye becomes the key sense organ also in the artistic domain. Generally speaking, mimetic theory in its various forms would have dominated reflection on art until the twentieth century with its avant-garde movements, according to the analysis presented by Danto (1983). In the philosophy of art history proposed by this American philosopher, art history would have ended—and art would have entered posthistory—once the eye has been stripped of its aesthetic primacy, i.e. once mimesis is no longer useful to explain art. The completion of mimesis and the end of the dominance of the eye make it possible to think anew the role of hand in art, as well as the concept of art itself. Some philosophers, such as phenomenologists or proponents of somaesthetics, have done so by reconsidering the concept of aesthetic experience and the traditional division between the higher and the lower senses. For these thinkers, aesthetic experience is fundamentally a bodily experience, not a purely intellectual one. They try to counteract the dominant Western conception of aesthetic experience as a purely mental reality.

## 2 The Lower Senses and Their Aesthetic Experience

We inherited from Plato the notion that the pre-eminent senses are sight and hearing, as he considered them the "intellectual" or higher senses. The so-called lower senses or secondary senses—taste, smell and touch—have been anaesthetized by Western civilization. They have been considered "primitive" and "unsophisticated" modes of artistic interaction to the extent that, according to Diaconu (2005), the main laws of Western aesthetics are "noli me tangere", "non olet" and "de gustibus non est disputandum". Even today, it is discussed whether art can effectively be created through using them, i.e. if attempts to turn activities such as gastronomy or perfumery into arts can be theoretically defended (Dutton 2009). The ephemerality of these kinds of "arts" that are oriented by these senses was already seen by Hegel to be an obstacle for a "universal" aesthetic experience. Hegel supported only those senses that somehow seem inherently receptive to the idealization, the "theoretical senses". He writes: "The sensuous aspect of art is related only to the two theoretical senses of sight and hearing, while smell, taste and touch remain excluded from the enjoyment of art" (Hegel 1975, p. 38). These three senses have to do with matter as such and its immediately sensible qualities, and so these senses cannot be related to artistic objects "which are meant to maintain themselves in their real independence and allow of no *purely* sensuous relationship" (Ibid.).

Among the lower senses, in traditional philosophical thinking, touch is in the lowest position. Plato's idea is that sight is the highest sense because it is the most distant and theoretical and touch the lowest because it is immediate. On the contrary, Aristotle considered touch to be the most universal of all: "without touch it is impossible to have any other sense; for every body that has soul in it must, as we have said, be capable of touch" (Aristotle, *De Anima*, 435ab).[2] Plato's idea was more influential, and thus Aristotle's position was defeated. This philosophical vanquishment has also an artistic counterpart. A. Gopnik has written: "Every other sense has an art to go with it: the eyes have art, the ears have music, even the nose and the tongue have perfume and gastronomy. But we don't train our hands to touch as we train our eyes to look or our ears to listen. Every now and again, someone comes up with a 'touch museum' or starts a program for the visually handicapped to experience art through their fingers. But such enterprises often have a hopeful, doomed feeling to them: they seem more willed than wanted" (Gopnik 2016).

---

[2]Aristotle defended as well that the hand is the tool of tools, the instrument that includes other instruments, and in this it is similar to the soul (Aristotle, *De Anima*, 432a). Aquinas comments on this: "The soul resembles the hand. The hand is the most perfect of organs, for it takes the place in man of all the organs given to other animals for purposes of defence or attack or covering. Man can provide all these needs for himself with his hands. And in the same way the soul in man takes the place of all the forms of being, so that through his soul a man is, in a way, all being or everything; his soul being able to assimilate all the forms of being—the intellect intelligible forms and the senses sensible forms". (Aquinas 1951, L.3, lect. 3, n. 790).

In Gopnik's statement, we perceive both the Western reduction of the aesthetic in general to the visual and auditory in particular, and the regret caused by the conviction that the art that is made to be touched—and haptic art in general—is actually some sort of vicarious, substitute art for those who cannot, for whatever reason, enjoy or experience "real" art. "Real" art would be the art made for the higher senses, which are what can really provide the "aesthetic experience" and allegedly fulfil that "real" purpose of art. The human hand, for that matter, could only provide a substitute for artistic experience, and only secondarily an aesthetic experience. Actually, when a work of art is said to have a tactile or haptic quality, what is actually meant is that it has a specific visual property.

Aesthetic experience has generally been considered a typically mental experience. It is supposed to be caused by a process related to the higher senses, which are "naturally" connected to the intellectual life. It has been regarded as something that refers to a particular state of mind, with its own qualities (disinterest, focused attention, fulfilment, etc.), which would be caused by the "contemplation" of certain realities that have some properties, and which cannot be identified with other specific mental states. This occurs rather paradigmatically in Kant's thought, where the judgment of taste, which judges something to be beautiful, is associated with a pleasurable experience and with a process of free play of cognitive faculties (imagination and understanding) not aimed at conceptual knowledge. It is an experience, if we can say so, that is "formally" cognitive, but not conceptual, based on a disinterested judgment. This idea of disinterest also dominates Schopenhauer's view, who focuses on the pleasure derived from the object itself, the pure contemplation of which saves us from the tyranny of the Will and, at the same time, allows us to apprehend the Platonic ideas; i.e. varying degrees of objectification of the Will (except in the case of music, which perceives the Will itself). Also in this case, the aesthetic experience is primarily an intellectual experience. As a form of "contemplation", the privileged sense of aesthetic experience is sight and, subsequently, hearing. The lower senses are thought to "intervene" and somehow pollute the idea. The hand transforms the idea by turning it into an "idea-hand" while, on the contrary, sight or hearing preserve the idea in its pristine purity. In parallel with the idea that creating a work of art is fundamentally an intellectual process, the aesthetic reception of the work of art is also considered purely contemplative. The body, in its various manifestations, would only be a means that would ultimately be dispensable. The ideal aesthetic experience would be that of a possible separate intellect that could experience things without using the body; the ideal creative process would be that of paintings (icons) that are said to be *acheiropoieta*, i.e. not made by human hands, in which the idea itself is supposed to be present in its original purity. Since this seems to be impossible, sight and hearing become acceptable *Ersätze*. In fact, throughout the history of art, we find reflections from different artists related to the judgment of the eye (Michelangelo), of hearing (Debussy), etc. usually used as revolutionary elements trying to counteract the primacy of intellectual principles. But we cannot find similar reflections on the "judgment of the hand" despite the clear intervention of the hand in the creation of artwork, in some way or another. Think about painting or about the many tiny nuances a violinist can create in music just by exercising different pressures on the strings or by changing the position of the hand on the bow. The overall intellectual conception of the work of art that has dominated Western Aesthetics makes it difficult to consider all these "handiworks" in a fair way. Among the few exceptions, the most important are perhaps the reflections developed by John Dewey in his work *Art as Experience*. In this text, Dewey complains about the theories that set aside art as a separate realm and holds that the task of the philosophy of art is "to restore the continuity between the refined and intensified forms of experience that are works of art and the everyday events, doings, and sufferings that are universally recognized to constitute experience" (Dewey 2005, p. 2), that is, "recovering the continuity of aesthetic experience with normal processes of living" (Ibid.: 9). For Dewey, the work of art is the device that operates in a way that makes our experiences so complete, full and total in their vitality as possible. From this conviction arise recent developments—such as Richard Schusterman's somaesthetics—which try to put the body in the foreground and recover Baumgarten's founding intuitions. In the phenomenological tradition, the works of M. Diaconu, in addition to analysing these philosophical foundations for a new paradigm of art—following the conversion of Aesthetics into an "Aisthetika" or theory of sensory perception developed by Gernot Böhme, Wolfgang Welsch and Martin Seel—, reveal how in avant-garde art these "secondary senses" are very present, to the extent that some important works are just a vindication of secondary senses as aesthetic senses. In this tradition—which, generally speaking, still gives preference to sight and hearing —there are several authors (e.g. Merleau-Ponty, Levinas, Henry, Irigaray or Marion), who vindicate the various senses in their conception of the human being and provide valuable insights for a new Aesthetic. Paradigmatically, Merleau-Ponty has reflected on the "cognitive character" of touching. He says: "when I touch my right hand with my left, my right hand, as an object, has the strange property of being able to feel too (…)". When I press my two hands together, it is not a matter of two sensations felt together as one perceives two objects placed side by side, but of an ambiguous set-up in which both hands can alternate the roles of 'touching' and being 'touched'. (…). In passing from one role to the other, I can identify the hand touched as the same one which will in a moment be touching. (…). The body catches itself from the outside engaged in a cognitive process; it tries to touch itself while being touched, and initiates 'a kind of reflection' which is sufficient to distinguish it from objects, of which I can indeed say that they 'touch' my body, but only when it is inert, and therefore without ever catching it unawares in its exploratory function" (Merleau-Ponty 2002, p. 106–107). As Diaconu has stated, this particular and rich "experiencing thinking" (*erfahrendes Denken*) goes through (*er-fährt,* in the Heideggerian meaning of the word) its object, instead of re-presenting it (*vor-stellen*) (Diaconu 2003, p. 4). Like the exploratory gaze of true vision, the "knowing touch" projects us outside our body through movement, a phenomenal component of tactile data. We can so perceive the patterning of tactile phenomena, just—as Merleau-Ponty says—as light exhibits the configuration of a visible surface. "It is not consciousness which touches or feels, but the hand, and the hand is, as Kant says, 'an outer brain of man'". While visual experience gives us the impression of an experience from nowhere, tactile experience is located on the surface of our body. "The body is borne towards tactile experience by all its surfaces and all its organs simultaneously, and carries with it a certain typical structure of the tactile 'world'" (Merleau-Ponty, 2002, pp. 367–370). For Merleau-Ponty, the hand is a specific place for the "experience of the body". For him, "it will perhaps be objected that the organization of our body is contingent, that we can 'conceive a man without hands, feet, head' and, *a fortiori* a sexless man, self-propagating by cutting or layering. But this is the case only if we take an abstract view of hands, feet, head or sexual apparatus, regarding them, that is, as fragments of matter, and ignoring their living function. Only, indeed, if we form an abstract notion of man in general, into which only the *Cogitatio* is allowed to enter. If, on the other hand, we conceive man in terms of his experience, that is to say, of his distinctive way of patterning the world, and if we reintegrate the 'organs' into the functional totality in which they play their part, a handless or sexless man is as inconceivable as one without the power of thought" (Merleau-Ponty 2002, p. 197). Having hands ("being-hands") is a "property" (*ídion*) of the human being, in the Aristotelian sense. For Aristotle, "a 'property' is a predicate which does not indicate the essence of a thing, but yet belongs to that thing alone, and is predicated convertibly of it" (Aristotle. *Topics* 1, c.5, 102a18–22, 282). Having hands is a normal part of the human experience. Specifically, having "living hands", as Aristotle stressed when he claimed that a severed hand, unable to fulfil its work, is only homonymously a hand, and not a real hand (Aristotle, *Metaphysics*, 1036, 30–33).

This idea of the human beings as possessing the property of having "living hands", which have a significant part in their relationship with reality, has a special importance in art. As it has been said, in the reflection on art, the hand has traditionally been associated almost exclusively with the process of creation, and it has been thought as a subsidiary element needed to carry out the idea possessed by the creative subject. The hand is seen as an organ that "translates" the idea into matter in the most faithful way possible (as a sort of perfect mimesis). But new forms of art intended to be received tactilely challenge this idea. Actual haptic art cannot be evaluated or appreciated by means of traditional aesthetic categories, but through categories conceived of in this new bodily way. This new way of thinking about the body in general and the hand in particular seems to be essential for understanding the changes that currently are challenging the essence of art. Examples of new forms of art, as, e.g. performance, show this overcoming of the visual-auditive paradigm of art.

## 3 To Touch or not to Touch

As we have remarked, the reception of the work of art has never been connected to tactile experience. Like the creation of the artwork, its reception is currently seen as a contemplative activity that is mainly concerned with grasping the idea it embodies. Prohibiting spectators to touch the works of art may have less to do with the interest in conserving an artwork (it is commonly argued that preventing physical contact will ensure this), and more to do with the a priori conception that the relationship with the artwork should be strictly contemplative. This is a deeply embedded idea arising from the very beginning of the modern conception of art: art is seen as a reality that maintains many of the features previously ascribed only to the divine. Our sincere relationship with the divine is purely contemplative, and so it must be with art (Castro 2017). But despite this allegedly purely intellectual relation to artworks, the fact is that many of today's so-called artworks arose in contexts of everyday life. Before being transformed into works of art—either converted or reduced to the status—it was not only common, but necessary to touch them. This is the case with works created for religious devotion. While maintaining their original context—e.g. churches, shrines and sanctuaries—they are touched and must be touched by the faithful as a sign of devotion and as a ritual that is a constituent part of a religious life; it is never just an "intellectual" process. But when they are treated as "pure" art, i.e. when they are presented in the context of an art exhibition, museum or any of the institutional frameworks of the art world, contemplative rules are imposed on them.

With the rise of visual arts, all modes of tactile aesthetics fell out of favour. There were some remarkable exceptions, like Goethe and Herder, who stressed the value of sculpture precisely in virtue of its tactile value (Classen 2007, pp. 901–902; 2012). It was not the general trend, however. The tendency to touch life-like sculptures in order to gain some kind of tactile experience of them was restricted to the religious sculptures, images and relics, which largely became what they are by being touched. Unlike old public or private collections and Cabinets of Curiosities, where artistic, scientific and religious oddities were lumped in together, the modern museum is born with new rules arising from the contemplative ideal inherited from the thought of that age. "This characteristically premodern mingling of spheres of knowledge was accompanied by a multisensory understanding of the cosmos according to which crucial information was transmitted and discernible through all sensory channels. If the museum was a little cosmos then it too could be regarded as constituting a multisensory tapestry of colours, textures, sounds and smells" (Classen 2007, p. 907). Changes in scientific practice and theory led to the consideration of sight as the only scientific sense of any value. With all the technical developments, no one needed to smell, touch or taste the material in order to know what it was. "As regards the museum, this sensory shift meant that allowing visitors close contact with museum pieces could no longer be justified by scientific values. The important thing in modernity was to see" (Ibid.). Due to the fact that more people began to visit museums, not only was there an increasing danger to the objects of art by visitors touching the objects, but there was also a sense that the vulgar touch of the masses was not like the contemplation of the elite. The "unwashed masses" were prone to destroy art by touching it; this is what all previous revolutionary movements had proven. From now on, museums of art are seen as contemplative spaces where the rule is "hands-off", breaking with the common practice. As Classen has stated "visitors to early museums were guided through the collection by a curator, just as guests might be guided through a private collection by a host. Allowing the visitors to touch the artefacts was an expected mark of courtesy on the part of the curator, who played the role of the host" (Classen 2007, p. 898).

Due to the current artistic developments, we can predict that this "hands-off" rule will not be a general norm in the museums of the future (Urist 2016). Even though we still live under the modern paradigm of "not touching", there are works that are intentionally made to circumvent or mock the rule of not touching the works of art (such as Duchamp's kinetic sculptures or his undecidable *Prière de toucher*). Paradoxically, they have also fallen under the institutional rule of "hands-off". Generally speaking, art is not interactive or, if I may say so, it is only intellectually interactive. In this paradigm, "permission" to touch the artwork is granted only to the experts (restorers, historians, curators, auctioneers, etc.). As a certain continuity of touch (*con-tact*) by the artist preserves the aura of the work of art, in order to maintain this sacred quality, the access thereto (*tact*) is restricted (*in-tact*). This constraint of the permission to touch brings out the association between touch and status (Candlin 2004, pp. 78–79), but at the same time between touch and brutality or primitivism. Unlike sight, something that is described by the scholastics as assuming the form of the object without violating it, touch has an invasive dimension. When touch is not wished for or allowed, it becomes violent *ipso facto* and forces as no other sense could do. The hand is especially controversial in this regard. The sense of touch manifests all over the skin, but the hand has anatomical peculiarities—which have led to many reflections on the relationship between the anatomy of the hand and brain development, or on human superiority due to the ability to manipulate things—that allow it to have a privileged role in the configuration of touch in the collective imagination. Actually, the touch of the hand is completely different from the touch of any other part of the body. The hand is itself an intentional organ, unlike, for example, the elbow or the knee.

This ambiguity was manifest in a recent exhibition at the MOMA. It was a re-enactment of "Imponderabilia", a piece originally performed by Marina Abramovic and a partner in the 1970s. A dancer re-enacting this piece was naked in a gallery entrance, facing a naked woman, as museum goers passed through the narrow space between them. In order to gain access to the next room, the visitor had to touch the artists in some way. One of the guests went a little too far and the artist said to the security guards: "This man is touching me" (La Rocco 2010). A distinction had arisen between touching that was demanded by the work and what the artist considered "inappropriate touching". The hand is always involved in this improper or "impure" touching.

Overall, this idea of impurity is mainly associated with touching above all the other senses. The previously mentioned fact that there are people that consider that touching a work of art causes it to deteriorate, also forms part of this view. To counteract this assessment, some authors are reflecting on the "patina", the surface tactile and visual quality caused by human contact (or weather) (Diaconu 2005). They are beginning to see it not as a destructive feature, but as a constructive trait. This aesthetic character emerges from the many traces left by different agents and, to some extent, is a part of the history of the object (Diaconu 2010, p. 318). Seen from a different point of view, it also provides a bit of a chronicle for the artwork. The patina, which in some contexts is considered an aspect that is part of the work of art (and sometimes of its own aura), raises a number of ontological problems concerning restoration, originality, etc. (Danto 2013, p. 53 ff.) It opens up a new aesthetic perspective that, if not entirely, at least in part has to do with the role of the hand in the reception of the work of art.

## 4 Conclusion

The vindication of the sense of touch in general and of the role of the hand in particular is having important consequences for Art theory and Aesthetics. Touching, to some extent, changes the being of things. Touching is a gesture saturated of meaning that can turn an object into a work of art if we understand art through new categories and expand the concept thereof. This claim of the active and passive element of touching and of the role of the hand as one of its specific and proper means forces us to rethink aesthetic experience and artistic activity from the point of view of the haptic. This new experience demands new aesthetic categories, yet to be thought through.

**Acknowledgement** The author is deeply indebted to Sudabee Lotfian for her help in translating and improving this paper.

## References

Aquinas, T. (1951). *Sentencia libri De anima.* Trans. Kenelm Foster, O.P. and Sylvester Humphries, O.P. New Haven: Yale University Press, 1951. http://dhspriory.org/thomas/DeAnima.htm#22 .
Aristotle. *Metaphysics*. Trans. W. D. Ross. The Internet Classics Archive, at http://classics.mit.edu/Aristotle/metaphysics.html.
Aristotle. *Topics*. Trans. W. A. Pickard-Cambridge. http://classics.mit.edu//Aristotle/topics.html.
Aristotle. *De Anima (On the soul)*. Trans J. A. Smith. http://psychclassics.yorku.ca/Aristotle/De-anima/de-anima3.htm.
Baumgarten, A. G. (2011). *Metaphysica*. Stuttgart Bad-Cannstaat: Fromman-Holzboog.
Candlin, F. (2004). Don't touch! Hands off! art, blindness and the conservation of expertise. *Body and Society, 10,* 71–90.
Castro, S. J. (2017). *Teología estética*. Salamanca: San Esteban.
Classen, C. (2007). Museum manners: The sensory life of the early museum. *Journal of Social History, 40,* 895–914.
Classen, C. (2012). *The deepest sense. A cultural history of touch*. Chicago: University of Illinois.
Danto, A. C. (1983). *The transfiguration of the commonplace*. Cambridge, Mass: Harvard University Press.
Danto, A. C. (2013). *What art is*. New Haven and London: Yale University Press.
Dutton, D. (2009). *The art instinct. Beauty, pleasure and human evolution*. New York: Bloomsbury Press.

Dewey, John. (2005). *Art as experience.* New York: The Berkley Publishing Group.
Diaconu, M. (2003). The rebellion of the "lower" senses: a phenomenological aesthetics of touch, smell, and taste. In C.-F. Cheunmg, I. Chvatik, et al. (Eds.), *Essays in celebration of the founding of the organization of phenomenological organizations.* www.o-p-o.net.
Diaconu, M. (2005). *Tasten, Riechen, Schmecken. Eine Ästhetik der anästhesierten Sinne.* Würzburg: Königshausen & Neumann.
Diaconu, M. (2010). Secondary senses. In H. R. Sepp, & L. Embree (Eds.), *Handbook of Phenomenological Aesthetics.* pp. 317–319, Dordrecht: Springer.
Dubost, J. (2014). Gefühl. In B. Cassin, (Ed.), *Dictionary of Untranslatables. A Philosophical Lexicon,* pp. 355–360, Princeton and Oxford: Princeton University Press.
Gopnik, A. (2016). Feel me. What the new science of touch says about ourselves. *The New Yorker.* May 16.
Hegel, G. W. F. (1975). *Aesthetics. Lectures on Fine Art.* Vol. I. Trans. T.M. Knox. Oxford: Clarendon, Press.
Kant, I. (1987). *Critique of Judgement, translated by Werner S.* Pluhar, Indianapolis: Cambridge Hackett Publishing Company.
La Rocco, C. (2010). Some at MoMA Show Forget 'Look but Don't Touch'. *New York Times.* April 16.
Shiner, L. (2001). *The invention of art. A cultural history.* Chicago and London: The University of Chicago Press.
Merleau-Ponty, M. (2002). *Phenomenology of perception.* Trans. Colin Smith, London & New York: Routledge.
Panofsky, E. (1968). *Idea.* New York: Harper & Row.
Urist, J. (2016). A new way to see art, *The Atlantic,* June 8.

# On the Role of the Hand in the Expression of Music

Marc Leman, Luc Nijs and Nicola Di Stefano

**Abstract** In diverse interaction processes that characterize music experience, the human hand can be seen as a mediator and facilitator for the brain's processing of musical expressive patterns. After a brief overview on the human expressive system for music, we consider gestures and hand articulations in music production and performance, focusing on hand dexterity and hand dystonia. Then, we discuss the role of the hand in music listening, conducting, and learning, showing that both in sound-generation and sound-accompaniment the hand mediates and facilitates action and perception in relation to musical expression. The recent use of technology in the domain of music is also considered throughout the chapter, with particular reference to sensing and motion technologies that allow users to control music parameters through hand–body movements. The hand can be considered as a co-articulated organ of the brain's action–perception machinery. Therefore, future research on hand and music should adopt a multiperspective approach that integrates different disciplines, from motor control to music performance and expression theories.

**Keywords** Hand · Gestures · Musical expression · Embodiment

## 1 Introduction

Interacting with music requires sophisticated motor skills in which hands seem to play a crucial role, both for sound-generation as well as for co-articulation and facilitation of expression (i.e., the timing, articulation, and intonation nuances of the musical structure). While the notion of action–perception is quite established as a

M. Leman · L. Nijs
IPEM—Musicology, Ghent University, Miriam Makebaplein 1, 9000 Ghent, Belgium

N. Di Stefano (✉)
Università Campus Bio-Medico di Roma, FAST—Institute of Philosophy of Scientific and Technological Practice, Via Alvaro Del Portillo 21, 00128 Rome, Italy
e-mail: n.distefano@unicampus.it

core concept of music understanding (see e.g., Leman 2007), it is less well established in terms of the specific role of the hand as a crucial effector in music interactions involving music playing, listening, and conducting. For example, while playing the piano, hands move rapidly over the keyboard, seemingly independently from each other. The motor system controls the hands to press the piano keys in the right place and at the right time, with precise articulation and adequate pressure of finger movements (Drake and Palmer 2000). But how is it possible to keep track of all those movements at the same time? In a similar way, hands are used to co-articulate the expression of musical sounds. Singers all over the world use their hands to support their vocal expression while singing dramatic stories (Davidson 2001). Why are they doing this? Listeners also tend to use their hands while listening to music, as if hands that co-articulate make the ephemeral experience a reality in space and time, as if the enactment of music facilitates its comprehension (Maes et al. 2014). Therefore, it is not exaggerated to say that hands play a central role in music interaction. This is most obvious during music playing, but even during listening, hands appear to be crucial effectors that facilitate the prediction of structural cues and their meaning (Sammler et al. 2013).

In this chapter, we review the specific role of the hand in musical expression. To fully comprehend this role, we consider hand movement from the perspective of a dynamic system, which implies that we study the hand as integrated part of the brain–body–environment interaction. We then discuss the role of the hand in view of an overall theory of musical expression and consider its role in music performance, listening and conducting.

## 2 Hand, Brain, and Music: A Dynamic System for Performing

According to El Koura and Singh (2003), the movement of the hand and fingers needed for interacting with music has 27 degrees of freedom: 4 in each finger, 3 for extension and flexion and one for abduction and adduction; 5 degrees of freedom in thumb, and 6 degrees of freedom for the rotation and translation of the wrist. Obviously, the most sophisticated hand dexterity is found in instrumental music playing, with the type of motor control varying according to the type of instrument. Depending on whether you play piano, double bass, or clarinet, you use hands and fingers in entirely different ways. How is it possible to control so many degrees of freedom? Apparently, no musician has to think about controlling hand muscles and finger parts.

Many researchers assume that the control of the hand and fingers is mediated by a small number of highly trained neuromuscular patterns, called motor primitives or muscle synergies, into which different types of muscle control can be grouped (d'Avella et al. 2003). In arm movements as well, muscles work in functional clusters that generally match the functional repertoire of the human arm (Gritsenko et al. 2016). Neuromuscular patterns thereby reduce movement complexity by reducing the dimensionality of available solutions. It is likely that similar neuromuscular patterns serve as building blocks to control fine-motor movements with a variety of timing and speed, such as while coordinating movement across fingers in the same and/or opposite direction. A small number of neuromuscular patterns thus allows for the control of a variety of movement repertoires for performing (Furuya and Altenmüller 2015).

However, neuromuscular patterns are also related to the player's intentions, in particular to their expressive execution. Firstly, trained neuromuscular patterns have to be associated with perceived audio patterns and visual scores (Wolpert et al. 2011), so that the perception of a chord scale, say G#7b5, selects the appropriate neuromuscular patterns needed to control the many degrees of freedom of the hand. Secondly, when the brain manages to make the right association between a perceived chord scale and the neuromuscular pattern to enact that chord scale, then that chord scale will be executed. Thanks to many rehearsals, the execution of that chord scale will be performed "automatically," i.e., without conscious control of all the fine details of how fingers move on the instrument. When that happens, we say that the chord scale resides in the musician's fingers. In reality, the musician's brain has constructed an association model that connects perception to action (the so-called inverse model) and realizes the action automatically during execution (the so-called forward prediction model) (Maes et al. 2014).

Playing by heart, when music resides in the fingers, requires a complex brain-machinery that supports hand dexterity in interaction with its musical environment. The brain–hand–environment system should always be kept in mind when considering music performance. Therefore, whatever we say about the hand of a musician should be considered from the viewpoint of a human brain in close interaction with its musical environment. Moreover, the hand's capacity as a tactile and haptic sensor of instruments, its role as precise working instrument for musical action, or its expressive display of musical intentions all work in parallel with the brain's ability to predict, control and express human interaction with the environment (MacDougall 1905; Cole 2013; Putz and Tuppek 1999).

The above considerations also suggest that the musician's hand can be seen as a mediator and facilitator for the brain's processing of musical expressive patterns. While generating musical expression involves dexterity to handle the instrument that realizes the transition from movement to sound, other types of music interactions such as listening, may require a lesser degree of dexterity—at least as far as finger dexterity is concerned. However, the hand's movement is also controlled by the brain's processing of musical expressive patterns available in the environment. In both sound-generation and sound-accompaniment, the hand and brain can be considered a single organ that mediates and facilitates action and perception in relation to musical expression. In other words, since hands and brains evolved together, they exist as a *reciprocal* unit that is implicative of "feedforward–feedback processes in which the hand and the brain form a dynamic system that reaches into the world" (Gallagher 2013).

## 3 The Human Expressive System for Music

A dynamic systems' viewpoint implies that the hand is not only a faithful and passive executor of the brain's intentions, neuromuscular commands, and prediction modeling, but also a capacitor for motion that fundamentally contributes to thinking and cognition (Goldin-Meadow 2003; Lundborg 2013; Radman 2013; Wilson 1998).

When people present new and advanced ideas about a problem they are working on, these ideas are often co-articulated by hand movements, as if notions that cannot yet be assimilated into language can be expressed through movements of the hands (Goldin-Meadow and Wagner 2005). In addition to that capacity of co-articulating, Cappuccio and Shepherd (2013) describe how specific gestures of pointing are integral to the phylogenetic and ontogenetic development of symbolic thinking (see also Tomassello 2010). Morett (2014), investigating the role of hand gestures in second language learning, found that they support fundamental cognitive processes related to communication, encoding, and recall. Overall, the hand seems to reflect our mind, expressing our innermost thoughts and wishes (Lundborg 2013). No wonder the hand has a privileged role in the expressive communication of emotions, feelings, ideas, and intentions (see also Lhommet and Marsella 2015). Hands help us learn, as in language learning (e.g., Goldin-Meadow 2014) or learning mathematics (e.g., Novack et al. 2014; Alibali and Nathan 2012). Hand gestures thus stimulate our cognitive ability (Goldin-Meadow 2014), facilitate learning (Kirschner 2002) and trigger our memory system (e.g., Rowe et al. 2008).

Moreover, hands that co-articulate our thinking also have the capacity to co-articulate our intentions and our perceptions. This is an important aspect in the understanding of how humans make sense of music. For example, when listening to music or dancing, humans typically use their hands to enact the sounds, to shape them and to move along with them, as if the hands co-articulate our inner musical experience (Leman 2016). When choreographies (i.e., learned movement sequences) of the hand are used, hand movements may resemble a form of conducting, especially when hand gestures seem to direct the musical performance that is heard. No wonder that in music education, hands are used when learning about music. A famous example is the "Guidonean hand," which is a sight-sing method for learning and singing based on the correspondence between the position of hand and fingers and musical schemas, i.e., sequences of tones (Reisenweaver 2012). The Curwen hand signs or the conducting gestures in solfege class are similar examples of this use of the hands.

Yet hand gesticulations that co-articulate thinking, intending, perceiving, enacting and so on, may seem coarse-grained compared to the fine-grained gestures needed to perform music on instruments. Obviously, playing a musical instrument is a domain *par excellence* in which the fine-motor dual functionality of the hand (i.e., as a working organ and a tool of action and intentionality) is needed and, therefore, most advanced. The rapid movements of the violinist's fingers, the complexity of a guitarist's grip, or the delicate touch of the pianist's fingers striking the keys, illustrate the highly sophisticated dexterity and sensitivity of human hand–brain–music capacity. In that context, it is of interest to consider a model of expressive musical communication that could serve as our overall dynamic systems perspective for understanding the role of the hand in dealing with music, and with musical expression in particular.

According to Leman (2016), musical communication can be understood as an interactive encounter, in which the enacting of the fluidity of a continuous stream of sound patterns plays a central role. By engaging brain processes that perform an alignment between hand and music (or musical expression), sound patterns get experienced in relation to these brain processes. More specifically, when the fluidity of musical expression is captured by hand and body gestures, the listener engages predictive, energetic, and affective processes so that sound patterns in turn get experienced as expressions of intentional, energetic and emotional states (either our own states, or states of the world onto which our own states can be projected). In that perspective, the alignment of hand gestures with musical expression can be seen as a method through which an enactment process can be realized. The basis of this type of enactive interaction is accuracy in timing. In responding to music, anticipation mechanisms (i.e., prediction processes involving forward models) are engaged to allow for a reverse of sound-motor sensing. According to Leman (2016), enaction of music (e.g., through hand movements) implies that the motor act can come first in order to anticipate the sound event in music, as if the sound is announced by the hand gesture in advance, or at the precise time that marks the end of hand gesture. Accordingly, the core of musical communication lies in the ability to turn very complex sound patterns into units that can be dealt with in terms of actions.

In short, drawing on corporeal expressive reflexes and on acquired repertoires of actions, sensory information is associated with actions and through associated energetic and affective states, sensory input is imbued with action goals and action directions. The intrinsic meaningfulness of the musical interaction (cf. Lesaffre et al. 2017) then lies in the fact that exchange of expression can be maintained for a while (i.e., as a homeostasis, or "equal state"). As such, a stable regime of interdependent processes (predictive, energetic, affective) is established, which sets a general condition of being exceptionally rewarding and therefore empowering for the person involved. Meaning, or sense-giving, is then conceived as the emergent outcome of an ongoing fluidic interaction, in which homeostasis is generated through synchronization and alignment of movements with sounds (Leman 2016).

In this theory of musical communication, every performer is a listener, and every listener is also a performer. The basis for both conditions is the construction of models that predict the information exchange in relation to the context in which it happens. During interaction with music, the information exchange has a dynamic character that is related to the causation of state changes. Hand movements play a crucial role in this enactment process because they have sufficient degrees of freedom to allow for a time-accurate alignment with, and prediction of, musical fluidity. Being able to align one's movements with musical expression has important consequences, of which reward is probably the foremost (Leman 2016).

Again, it can be argued that the hand plays an important role because it allows for the expression of gestures whose anticipation can be rewarded by the musical environment, for example when the anticipation expresses a particular time-critical outcome that is then satisfied by the music exactly at the moment indicated by the hands' movement.

The human expressive system entails not only the sensitivity to affordances in the environment (i.e., elements in the environment that provoke an expressive response such as a child screaming, a dog barking) but also the capacity of having responses (including anticipations) to those affordances (singing a lullaby, prattle). As it comprises a reflexive urge to respond to sound by means of body movements, or by means of expressive sounds, the expressive system is biologically determined. As it comprises a learned control of reflexes in view of expressive gestures, it is also culturally determined. According to this theory (Leman 2016), the architecture of the expressive system therefore encompasses different components such as: musical affordances involving a set of expressive cues embedded in sonic cues; sensori-motor mechanisms, based on associative and predictive schemes that constitute the expressive gesturing repertoire; and a motivational system, which is based on rewards that stimulate an autotelic and sense-giving dimension of musical engagement. The term "expressive" can thereby be defined as the quality of a movement or sound that stirs the interaction, due to the fact that it provokes a response (body movement, sound) that is also expressive.

As the hand is a major component in music interaction, it is assumed that hand gesture patterns reveal information about interaction states, and thus about the predictive, energetic and emotional processes that support interactions. For example, patterns in hand gestures might reveal information about energetic and affective states (e.g., happy vs. sad) attributed to a piece of music, the structures that are perceived in it (e.g., binary vs. ternary) or what is predicted to come next (e.g., dynamic change) (see Maes et al. 2014).

Gestural responses to music and their basis in a latent repertoire conferred by biology are the focus of a current debate (see Byrne and Cochet 2017). As this question is driving much of today's work regarding hand-music research, we cannot offer an exhaustive overview of ongoing work. Instead, we provide some illustrative examples of research that focuses on the hand in different forms of musical interaction: music performance, music listening, conducting and dance.

## 4 The Hand in Music Performance

The privileged role of the hand in the expression of music is perhaps most evident from the perspective of music performance and sound production. Playing a musical instrument in an expressive way needs great spatiotemporal precision and highly advanced sensory functions. Music performance thus requires sophisticated actions (including fingering, tonguing; i.e., techniques of using the fingers and the tongue in articulating wind-instruments) that encode the musical intentions (i.e., tones and their expressive design) into sound patterns. Through bodily effectors such as motions of shoulders, arms, hands, and fingers that control the musical instrument, sound patterns produced by those instruments are endowed with expression (Leman 2007). That expressive motility, reflected in sound, touches in turn the aesthetic and affective curiosity of the listener (to be considered in the next section).

For many years, hands and fingerings have been a topic of study in music research. In recent years, methods for measuring hand movement have been refined thanks to technological developments (e.g., motion capture systems, sensors), allowing an in-depth view of fine-grained movement data. Research topics that have received much interest are hand *dexterity* and hand *dystonia*. Hand dexterity refers to hand skills, while hand dystonia refers to abnormal muscle tone, often the result of over-trained dexterity.

## 4.1 Dexterity

One aspect of hand dexterity is concerned with *hand preference*. Findings from different studies suggest that left or mixed-handedness should be favored among instrumentalists due to the demand for highly accurate, rapid motor coordination. For example, based on the fact that left-handed people have better motor control of their non-preferred hand than do right-handers (Peters 1980; Kilshaw and Annett 1983), Annett (1985) speculated that the highly skillful playing of musical instruments might be more difficult among strong right-handers than among weak right-handers. In line with these findings, Christman (1993) found that left or mixed handers are more frequent among those musicians who play instruments requiring greater bimanual coordination. Aggleton et al. (1994) found a significant increase of left or mixed handers among musician compared to a control group.

The second aspect of dexterity is skills transfer due to *plastic adaptations* of the nervous system to sensorimotor training (Altenmüller and Furuya 2016). For example, Chang et al. (2014) examined the effect of music training on unimanual and bimanual responses adopting a simple tapping task and different types of participants (pianists, singers, and non-musicians). They found that non-musicians, compared to musicians, had a longer tapping response to visual cues. Interestingly, they did not find significant differences between pianists and singers, suggesting that music learning, even when it does not involve motor abilities, can improve bimanual coordination. Spilka et al. (2010) found that musicians are also generally better performers than non-musicians at imitating unfamiliar and complex manual gestures. This seems to confirm that instrumental training deeply affects hand dexterity and sensibility in different domains too. The advantage is linked to the ability to reproduce fine-motor components of model gestures, as musician's hands have been trained more on fine skills during the instrumental training.

A third aspect of dexterity is related to *fine-motor movements*. This involves control of motor primitives as well as the development and use of (inverse and forward) prediction models that facilitate automatisms of rapid fingering. Such automatisms are crucial in performance (Metcalf et al. 2014). Seen from this perspective, different studies have focused on the technical aspects of using the hands in piano playing, investigating octave strikes (Furuya and Kinoshita 2008; Furuya et al. 2011), fingertip movements (Dalla Bella and Palmer 2011; Goebl and Palmer 2013) or wrist motion (Hadjakos and Waloschek 2014; Oikawa et al. 2011). These studies point out that there exist skill level-dependent differences in movement strategies between experts and novices, such as lower energetic costs, utilizing a greater number of degrees of freedom as a function of musical expertise. With regard to expression in relation to the hands' degrees of freedom, Davidson (2007) noticed that pianists displayed more expressive movements when the hand was free or in *rubato* sections where there is more time available to show the movement quality. A fourth aspect of dexterity concerns *dynamic brain–hand interactions*. In recent years, new ways of performing have emerged in which different kinds of sensing technologies allow users to control music parameters through body movement, muscle tone, and posture. Consequently, the search for intuitive mappings between music and movement has led to a vast number of studies that address the music-movement connection considering hands as a musical instrument themselves. In that context, the role of performing musicians in scientific research on hand movement has been significant. For example, Michel Waisvisz is well-known as the developer and virtuoso of "The Hands," a performance interface based on hand-mounted sensors. In an interview, he stated: "I see the hand as a part of the brain, not as a lower instrument of the brain. Of course, you can see a hand as a transmitter and sensor, but in the consciousness of the performance, the hand is the brain. You can't say that its precision is surpassed or even equaled by computers because we simply don't know what we control in detail when we play an instrument. Every instrumentalist can tell you that in the instrumental learning process, there are hours of meticulous motor memorization of timing and intonation, but the thing called music finally comes out as something on top of that" (Krefeld and Waisvisz 1990).

In view of dexterity and the dynamic relationship between brain and hand, the design of straightforward mappings between hand gestures and sound gestures is central to every musical instruments' maker (Hunt et al. 2000). Capturing the subtle sensitivity and dexterity of the hand is a major challenge, especially in electronic musical instruments where mappings are often arbitrary, or where audio effects are delayed due to processing. For example, a hand gesture, whose sound-producing cue is picked up at a particular spatiotemporal instance may not lead to the expected output when the outcome occurs 50 ms later than expected. Similar delays are typically due to the processing of the cue and to the audio processing needed to generate the sound. Such delays may be extremely annoying in expressive interactions. Solutions would require active responses from the instrument, that is, responses that anticipate human hand dexterity. An example of such a system is PLXTRM (Vets et al. 2017), an extended guitar that predicts about 50 ms ahead in time whether a guitar string will be plucked. That anticipation is then used to prepare proper audio effects whose timing better corresponds with the hand and finger gestures and their corresponding sound effect expectations. Instruments with additional features that increase the traditional instrument's functionality are called augmented instruments: they all deal with the hand–brain–music dynamics, time accuracy, and expressiveness.

## 4.2 Dystonia

Altenmüller and Jabusch (2009) define musician's dystonia as a task-specific movement disorder, resulting from overuse of the neuromuscular system and which manifests itself as a loss of voluntary motor control in extensively trained movements. It may be accompanied by tremor, acute and chronic pain. In many cases, dystonia terminates the careers of affected musicians.

Dystonia is not limited to the hand. It may also affect brass player's lips. Typically, dystonia develops in musicians who practice extensively for many hours a day for many years. Through training, musicians hope to mold the innate neuromuscular constraints into new neuromuscular constraints that allow multiple fingers to move with differing timing, at speeds that mark the virtuoso musician (Furuya and Altenmüller 2015). Typical symptoms of hand dystonia are involuntary hyperflexion of the finger joints and/or excessive cramping of finger muscles, which deteriorate spatiotemporal accuracy of finger movements and impair hand dexterity. Recent work in neuroscience focuses on the clinical effects of different intervention techniques on musician's dystonia, e.g., sensory and motor retraining, transcranial stimulation, injection of botulinum toxin, and medications (Furuya and Altenmüller 2015).

In short, the musician's hand is extremely capable of highly sophisticated movements needed for encoding human expression in sounds. These movements are controlled by neuromuscular patterns and predictive models governed by the brain. The process of acquisition of those patterns and models, however, is not without risks. Over-stimulation may lead to disorders, such as hand dystonia that terminates virtuosity and dexterity.

## 5 The Hand in Music Listening

Listening to music is a multimodal experience in which brain and hand are intimately connected (Patel 2014; Leman and Maes 2014). Neuroscientific studies provide evidence of a tight auditory-motor neural circuit on which this experience may be based. These studies point to the pathway linking the primary auditory cortex to the posterior parietal cortex and from there to the motor areas of the frontal cortex (Zatorre et al. 2007). Evidence for a perception-action linkage during musical activities comes from brain imaging studies in trained musicians (Bangert et al. 2006; Habib and Besson 2009).

Recent studies in music perception point to the role of hand-related movements in music perception (Manning and Schutz 2013; Maes et al. 2014). However, as noted, the hand-related movements of the listener require less sophisticated finger movements than the hand-related movements of the musician. Instead, the focus is more on hand-arm movements, and simple finger movements (such as needed for tapping) and time accuracy, rather than on hand-finger movements (independency of finger movements) and spatial accuracy. Given these constraints, Maes et al. (2014) argue that music-driven gestures can facilitate music perception because the movements obey forward models that may in turn prime auditory expectations. Komeilipoor et al. (2015) show that the regularity of repetitive finger movements performed while listening to consonant or dissonant sounds is influenced by the degree of consonance of the sound presented. Di Stefano et al. (2017) show that different stimuli, i.e., consonant and dissonant, evoke different motor behaviors in children while playing with a musical toy.

Overall, bodily responses to music listening involve hand-related responses in different ways. Low-level synchronizations are the hand–brain's reactions to acoustical stimuli. Tapping, for example, occurs when a subject spontaneously synchronizes his/her finger movements to the perceived rhythm (see Repp 2005, for a review). Research on these kinds of synchronizations has evidenced the priority of auditory-motor over visual-motor coupling among humans (Thaut et al. 1999; Repp 2003; Patel et al. 2005), thus providing further proof of the above-mentioned linkage between acoustic perception and motor entrainment.

Higher-level synchronizations are motor actions that occur when the listener/musician aims at expressing or conveying the emotional aspect of music. Gestures, for example, are bodily analogies to sonic properties related to expressiveness and emotions (Leman 2007). Recent studies have begun to consider gestures, and especially hand, arm, and head movements, as an essential aspect of musical performance (Ben-Tal 2012; e.g., Simones et al. 2015). In a study by Amelynck et al. (2014), listeners were asked to participate interacting spontaneously to music. The log-speed of the arm-hand movements correlated with the musical energy. Moreover, it was shown that expressive hand movements did not differ much among listeners. Focusing on the listeners' average hand movement and three characteristic movements that capture the variance among their hand movements above that average, it was possible to model the individual hand movements of each listener with high accuracy. The study suggests that despite the high degree of freedom in hand-arm effectors, movements that go along with the expressive fluidity of the music are in fact low dimensional. This finding of low-dimensionality may indicate that expression in music is in fact low dimensional and, therefore, through alignment, hand-arm movements tend to express that dimensionality. It should be noted that in the study of Amelynck et al. (2014) the hand was measured with a single marker. This reminds us that in studies of listening, a hand is often conceived more as a hand-arm unit rather than a hand-finger unit.

## 6 The Hand in Conducting

One of the most typical musical activities in which the hands play a crucial role is conducting a music ensemble or orchestra. Hands can indicate expressive intentions (e.g., fluid directional arcs), accompany characteristic bodily sways and articulations, control the musical structure (e.g., tempo, dynamics), or point to purely technical aspects (e.g., way of attacking a note in the wind section, tone quality). Whether leading small or big ensembles (e.g., symphony orchestra, woodwind band), the conductor's hands communicate important information to coordinate multiple players, manage stylistic requirements and inspire musicians. Sometimes one hand is extended with a baton to invoke temporal precision and efficiency, while the other hand indicates the mood through different positions (e.g., clenched fist for maestoso, little finger raised for delicacy, relaxed flowing movement for cantabile; McElheran 1989). In conventional accounts of conducting, very often the right hand is said to be responsible for time-beating gestures, while the left hand is mostly free from time-beating (unless mirroring or assisting the right hand) and used for cueing, indicating dynamics and style, aiding in tempo changes, controlling the balance. In any case, at least one hand will exploit its degrees of freedom in function of expressive communication to the musician, or to the audience. Using a baton or not, the conductor's gestures must be first and always meaningful in terms of music. Arm and hand gestures thereby provide more relevant information than faces, although faces hold important affective cues that go along with hands (Wöllner 2008; Fuelberth 2004). Conductors' (hand) gestures have also come under the scrutiny of researchers investigating how musical intentions are reflected in these gestures (e.g., Luck et al. 2010).

Utilizing motion technologies, it is possible to create interactive environments that allow a wide public to engage in the act of conducting. For example, Nakra et al. (2009) designed *UBS Virtual Maestro*, an interactive conducting system that provides the experience of orchestral conducting to the general public attending a classical music concert. Borchers et al. (2004) built the *Personal Orchestra*, an interactive exhibit that allows users to control the playback speed and volume of an audio/video recording of the Vienna Philharmonic by using Buchla "Lightning" batons (This system is continuously displayed at the *Haus der Musik* in Vienna). Another similar system, called *You're the Conductor*, featuring footage from the Boston Pops Orchestra, invites children to conduct an audio and video recording of an orchestra without prior musical experience. This system was on display at the Boston Children's Museum from 2003 to 2008 and toured to Children's Museums around the USA from 2004 to 2006 (Lee et al. 2004). Ways to model robot conducting are also considered in Amelynck et al. (in print).

Besides technological developments, there are also new ways of conducting that support and facilitate group improvisation in which participants are not required to have previously acquired improvisation experience or skills. For example, Zappa

(1973) used specific hand gestures to give his band members cues to create certain sound effects and even to give directions to the audience (Minors 2012). Although these hand gestures followed a tradition of musicians like Duke Ellington or Charles Mingus, Zappa used them in a unique way. A composition that had been scored and rehearsed in a specific way could be drastically changed by the spontaneous implementation of specific hand signals (Carr and Hand 2008). Another example is Sound Painting, in which hand gestures constitute a "comprehensive multidisciplinary sign language for creating live composition from structured improvisation" (Thompson 2010). The Sound painter stands in front of a group of musicians, signs a phrase to the group and then composes with the responses. Inspired by this way of making music, a similar system (VOPA or "Vocal Painting") was designed for singers. The signs are easy to understand and can be used to provoke musical expression within basic parameters. There are currently 75 VOPA signs and new signs continue to be developed. However, the first five signs are sufficient to achieve significant improvement in musical expression with any choir or vocal group.

## 7 The Hand in Music Learning

A final word should be said about the hand in music learning. The pedagogical approach of the hand in music playing has existed for many centuries. The so-called cheironomy, or doctrine of hand signs in music, is related to the neumes, or signs that served as a guide for singers who knew the melodies more or less by heart, or for the choir leader who may have interpreted them to the singers by appropriate movements of the hand. The neumes supposedly have their origin in the accentuation signs of Greek and Roman literature (Strayer 2013). But this use of the hand as a mnemonic device was also known in Indian (vedic), Tibetan, Japanese and Korean chants (Kaufman 1967). In the Western tradition, this use of the hand was extensively developed by Guido of Arezzo, an Italian medieval music theorist who introduced solmization. The "Guidonian hand" was a sight-sing method for learning and singing based on the connection of syllables to pitch and on the use of the hand as memory cue. The basic idea of his method is the correspondence between the position of hand and fingers and musical sequence of tones. Each part of the hand represents a note within the hexachord system. As such, it allowed singers to visualize where the semitone of the hexachords was, which was the fundamental element for correct tuning and singing (Reisenweaver 2012). The idea of using the hand and syllables as a memory cue for singing was also adopted by Curwen, who added a series of specific hand signs to Sarah Glover's syllables "to individualize" them and to aid in their memorization (Zinar 1983). Zoltán Kodály, a Hungarian music scholar and educator, made hand signs internationally recognized as a valuable pedagogical tool. However, despite the claims of many educators who use the Kodály hand signs, a scarcity of research has addressed the effectiveness of these signs in music learning. Several studies have addressed sight-singing, pitch accuracy or pitch discrimination skills by comparing groups that used the Kodály hand signs with groups that did not. For the most part these studies did not find statistically significant differences (for a short overview, see McClung 2008 and Reifinger 2012). Most of them involved short-term experiences with movable solfège syllables coupled with the use of hand signs. Steeves (1985) found that the use of hand signs supports faster and more correct identification of melodic intervals. Also, research indicates that showing pitch by gesture improves students' melodic progression (Langness 1997; Mueller 1993; Phillips 1992).

Other research addresses the possibility of (hand) gestures to provide motor sensations that can assist in the development of sound sensations, thereby improving technical aspects of singing such as intonation, tone quality and articulation. For example, Liao and Davidson (2007) showed that the shape of hand gestures can affect tone quality. In line with these findings, Nafisi (2013) found different types of gestures among which are hand gestures that visualize actual internal physiological mechanisms related to the singing process (physiological gestures), that illustrate singing metaphors, imagery and/or acoustic phenomena (sensation related gestures) and that give visible form to musical phenomena (musical gestures). Research on its effectiveness has over all been limited. However, in recent years the number of studies that address the hand in music is growing and a wider array of music-learning-related aspects is being investigated.

## 8 Conclusion

Throughout the history of mankind, hand sensitivity and dexterity have played an important role in the development of skills and crafts that have shaped the environment. In this chapter, we considered the hand from the viewpoint of a co-articulated organ of the brain's action–perception machinery. The brain's role in the encoding and decoding of expression is central as a prediction engine containing neuromuscular patterns and time-accurate models of motor control. Hand-arm articulations and hand-finger articulations therefore provide the backbone of any musical interaction, be it as performers, listeners or conductors. Studies on hands and music have focused primarily on motor control, action theory, or music performance and expression. The different approaches and viewpoints suggest that our understanding of the role of the hand in music interaction can advance by adopting an overall dynamic-systems approach, thus embracing the idea that the hand–brain–music unit comprises a networked system of units whose mutual interaction may result in highly sophisticated skills.

# References

Aggleton, J. P., Kentridge, R. W., & Good, J. M. M. (1994). Handedness and musical ability: A study of professional orchestral players, composers, and choir members. *Psychology of Music, 22*(2), 148–156.

Alibali, M. W., & Nathan, J. M. (2012). Embodiment in mathematics teaching and learning: Evidence from learners' and teachers' gestures. *Journal of the Learning Sciences, 21*(2), 247–286.

Altenmüller, E., & Furuya, S. (2016). Brain plasticity and the concept of metaplasticity in skilled musicians. In J. Laczko & M. L. Latash (Eds.), *Progress in motor control* (pp. 197–208). Basel, CH: Springer International Publishing.

Altenmüller, E., & Jabusch, H. -C. (2009). Focal hand dystonia in musicians: Phenomenology, etiology, and psychological trigger factors. *Journal of Hand Therapy, 22*(2), 144–155.

Amelynck, D., Maes, P.-J., Martens, J.-P., & Leman, M. (2014). Expressive body movement responses to music are coherent, consistent, and low dimensional. *IEEE Transactions on Cybernetics, 44*(12), 2288–2301.

Annett, M. (1985). *Left, right, hand and brain: The right shift theory*. Hillsdale, NJ: Erlbaum.

Bangert, M., Peschel, T., Schlaug, G., Rotte, M., Drescher, D., Hinrichs, H., Heinze, H.J. & Altenmueller, E. (2006). Shared networks for auditory and motor processing in professional pianists: Evidence from fMRI conjunction. *Neuroimage, 30*, 917–926.

Ben-Tal, O. (2012). Characterising musical gestures. *Musicae Scientiae, 16*(3), 247–261.

Borchers, J., Lee, E., Samminger, W., & Mühlhäuser, M. (2004). Personal orchestra: A real-time audio/video system for interactive conducting. *ACM Multimedia Systems Journal Special Issue on Multimedia Software Engineering, 9*(5), 458–465.

Byrne, R. W., & Cochet, H. (2017). Where have all the (ape) gestures gone? *Psychonomic Bulletin and Review, 24*(1), 68–71.

Cappuccio, M., & Shepherd, S.V. (2013). Pointing hand: Joint attention and embodied symbols. In Z. Radman (Ed.), *The hand, an organ of the mind. What the manual tells the mental*. Cambridge, MA: The MIT Press.

Carr, P., & Hand, R. J. (2008). Twist 'n frugg in an arrogant gesture: Frank Zappa and the Musical-Theatrical Gesture. *Popular Musicology Online* (March 2008). http://www.popular-musicology-online.com/issues/05/carr.html .

Chang, X., Wang, P., Zhang, Q., Feng, X., Zhang, C., & Zhou, P. (2014). The effect of music training on unimanual and bimanual responses. *Musicae Scientiae, 18*(4), 464–472.

Christman, S. (1993). Handedness in musicians: Bimanual constraints on performance. *Brain and Cognition, 22*(2), 266–272.

Cole, J. (2013). Capable of whatever man's ingenuity suggests: Agency, deafferentation, and the control of movement. In Z. Radman (Ed.), *The hand, an organ of the mind. What the manual tells the mental*. Cambridge, MA: The MIT Press.

d'Avella, A., Saltiel, P., & Bizzi, E. (2003). Combinations of muscle synergies in the construction of a natural motor behavior. *Nature Neuroscience, 6*(3), 300–308.

Dalla Bella, S., & Palmer, C. (2011). Rate effects on timing, key velocity, and finger kinematics in piano performance. *PLoS ONE, 6*(6), e20518.

Davidson, J. W. (2001). The role of the body in the production and perception of solo vocal performance: A case study of Annie Lennox. *Musicae Scientiae, 5*(2), 235–256.

Davidson, J. W. (2007). Qualitative insights into the use of expressive body movement in solo piano performance: A case study approach. *Psychology of Music, 35*(3), 381–401.

Di Stefano, N., Focaroli, V., Giuliani, A., Formica, D., Taffoni, F., & Keller, F. (2017). A new research method to test auditory preferences in young listeners: Results from a consonance versus dissonance perception study, *Psychology of Music, 45*(5), 699–712.

Drake, C., & Palmer, C. (2000). Skill acquisition in music performance: Relations between planning and temporal control. *Cognition, 74*(1), 1–32.

ElKoura, G., & Singh, K. (2003). Handrix: Animating the human hand. In *Proceedings of the 2003 ACM SIGGRAPH/Eurographics symposium on Computer Animation* (pp. 110–119). Eurographics Association.

Fuelberth, R. J. V. (2004). The effect of various left hand conducting gestures on perceptions of anticipated vocal tension in singers. *International Journal of Research in Choral Singing, 2*(1), 27–38.

Furuya, S., & Altenmüller, E. (2015). Acquisition and reacquisition of motor coordination in musicians. *Annals of the New York Academy of Sciences, 1337*(1), 118–124.

Furuya, S., & Kinoshita, H. (2008) Expertise-dependent modulation of muscular and non-muscular torques in multi-joint arm movements during piano keystroke. *Neuroscience, 156*, 390–402.

Furuya, S., Goda, T., Katayose, H., Miwa, H., & Nagata, N. (2011). Distinct inter-joint coordination during fast alternate keystrokes in pianists with superior skill. *Frontiers in Human Neuroscience, 5*, 1–13.

Gallagher, S. (2013). The enactive hand. In Z. Radman (Ed.), *The hand, an organ of the mind. What the manual tells the mental.* Cambridge, MA: The MIT Press.

Goebl, W., & Palmer, C. (2013). Temporal control and hand movement efficiency in skilled music performance. *PLoS ONE, 8*(1), e50901.

Goldin-Meadow, S. (2003). *Hearing gesture: How our hands help us think.* Cambridge, MA: Harvard University Press.

Goldin-Meadow, S. (2014). Widening the lens: What the manual modality reveals about language, learning and cognition. *Philosophical Transactions of the Royal Society B, 369*(1651), 20130295.

Goldin-Meadow, S., & Wagner, S. M. (2005). How our hands help us learn. *Trends in Cognitive Sciences, 9*(5), 234–241.

Gritsenko, V., Hardesty, R. L., Boots, M. T., & Yakovenko, S. (2016). Biomechanical constraints underlying motor primitives derived from the musculoskeletal anatomy of the human arm. *PLoS ONE, 11*(10), e0164050.

Habib, M., Besson, M. (2009). What do music training and musical experience teach us about brain plasticity? *Music Perception: An Interdisciplinary Journal, 26*(3), 279–285.

Hadjakos, A., & Waloschek, S. (2014). VIP: Controlling the sound of a piano with wrist-worn inertial sensors. In *International Conference on New Interfaces for Musical Expression, Keyboard Salon Workshop* (Vol. 30).

Hunt, A., Wanderley, M. M., & Kirk, R. (2000). Towards a model for instrumental mapping in expert musical interaction. In *Proceedings of the 2000 International Computer Music Conference*, ICMC, Berlin, Germany.

Kaufmann, W. (1967). The mudras in Samav-edic chant and their probable relationship to the go-on hakase of the Shomyo of Japan. *Ethnomusicology, 11*, 161–169.

Kilshaw, D., & Annett, M. (1983). Right- and left-hand skill I: Effects of age, sex and hand preference showing superior skill in left-handers. *British Journal of Psychology, 74*, 253–268.

Kirschner, P. A. (2002). Cognitive load theory: Implications of cognitive load theory on the design of learning. *Learning and Instruction, 12*, 1–10.

Komeilipoor, N., Rodger, M. W. M., Craig, C. M., & Cesari, P. (2015). (Dis-) Harmony in movement: Effects of musical dissonance on movement timing and form. *Experimental Brain Research, 233*, 1585–1595.

Krefeld, V., & Waisvisz, M. (1990). The hand in the web: An interview with Michel Waisvisz. *Computer Music Journal, 14*(2), 28–33.

Langness, A. P. (1997). Helping children's voices develop in general music education. In L. Thurman & G. Welch (Eds.), *Body mind and voice: Foundations of voice education* (pp. 571–581). Minneapolis, MN: Voice Care Network.

Lee, E., Nakra, T.M., & Borchers, J. (2004). You're the conductor: A realistic interactive conducting system for children. *Proceedings of the 2004 conference on new interfaces for musical expression*, 68–73.

Leman, M. (2007). *Embodied music cognition and mediation technology*. Cambridge, MA: The MIT Press.

Leman, M. (2016). *The expressive moment: How interaction (with music) shapes human empowerment*. Cambridge, MA: The MIT Press.

Leman, M., & Maes, P.-J. (2014). Music perception and embodied music cognition. In L. Shapiro (Ed.), *The routledge handbook of embodied cognition* (pp. 81–89). New York, NY: Routledge.

Lesaffre, M., Maes, P. J., & Leman, M. (Eds.). (2017). *The Routledge companion to embodied music interaction*. London: Routledge.

Lhommet, M., & Marsella, S. (2015). Expressing emotion through posture and gesture. In R. Calvo, S. D'Mello, J. Gratch, & A. Kappas (Eds.), *Handbook of affective computing* (pp. 273–285). New York: Oxford University Press.

Liao, M. Y., & Davidson, J. (2007). The use of gesture techniques in children's singing. *International Journal of Music Education, 25*(1), 82–96.

Luck, G., Toiviainen, P., & Thompson, M. R. (2010). Perception of expression in conductors' gestures: A continuous response study. *Music Perception, 28*(1), 44–57.

Lundborg, G. (2013). *The hand and the brain: From Lucy's thumb to the thought-controlled robotic hand*. New York, NY: Springer Science & Business Media.

MacDougall, R. (1905). The significance of the human hand in the evolution of mind. *The American Journal of Psychology, 16*(2), 232–242.

Maes, P.-J., Leman, M., Palmer, C., & Wanderley, M. M. (2014a). Action-based effects on music perception. *Frontiers in Psychology, 4*(1008), 1–14.

Maes, P.-J., Van Dyck, E., Lesaffre, M., Leman, M., & Kroonenberg, P. (2014b). The coupling of action and perception in musical meaning formation. *Music Perception, 32*(1), 67–84.

Manning, F., & Schutz, M. (2013). "Moving to the beat" improves timing perception. *Psychonomic Bulletin & Review, 20*(6), 1133–1139.

McClung, A. C. (2008). Sight-singing scores of high school choristers with extensive training in movable solfège syllables and Curwen hand signs. *Journal of Research in Music Education, 56*(3), 255–266.

McElheran, B. (1989). *Conducting technique: For beginners and professionals*. New York, NY: Oxford University Press.

Metcalf, C. D., Irvine, T. A., Sims, J. L., Wang, Y. L., Su, A. W., & Norris, D. O. (2014). Complex hand dexterity: A review of biomechanical methods for measuring musical performance. *Frontiers in Psychology, 5*, 414.

Minors, H. (2012). Music and movement in dialogue: Exploring gesture in soundpainting. *Les Cahiers de la Société québécoise de recherche en musique, 131*(2), 87–96.

Morett, L. M. (2014). When hands speak louder than words: The role of gesture in the communication, encoding, and recall of words in a novel second language. *The Modern Language Journal, 98*(3), 834–853.

Mueller, A. K. (1993). *The effect of movement-based instruction on the melodic perception of primary-age general music students*. Unpublished doctoral dissertation, Arizona State University, USA.

Nafisi, J. (2013). Gesture and body-movement as teaching and learning cools in the classical voice lesson. A survey into current practice. *British Journal of Music Education, 30*(3), 347–367.

Nakra, T. M., Ivanov, Y., Smaragdis, P., & Ault, C. (2009). The UBS virtual maestro: An interactive conducting system. In N. Zahler, & R. Dannenberg (Eds.). *Proceedings of the 2009 Conference on New Interfaces for Musical Expression* (pp. 250–255). Pittsburgh, PA: School of Music, Carnegie Mellon University.

Novack, M. A., Congdon, E. L., Hemani-Lopez, N., & Goldin-Meadow, S. (2014). From action to abstraction: Using the hands to learn math. *Psychological Science, 25,* 903–910.

Oikawa, N., Tsubota, S., Chikenji, T., Chin, G., & Aoki, M. (2011). Wrist positioning and muscle activities in the wrist extensor and flexor during piano playing. *Hong Kong Journal of Occupational Therapy, 21*(1), 41–46.

Patel, A. D. (2014). The evolutionary biology of musical rhythm: Was Darwin wrong? *PLoS Biology, 12*(3), e1001821.

Patel, A. D., Iversen, J. R., Chen, Y., & Repp, B. H. (2005). The influence of metricality and modality on synchronization with a beat. *Experimental Brain Research, 163*(2), 226–238.

Peters, M. (1980). Why the preferred hand taps more quickly than the non-preferred hand: Three experiments on handedness. *Canadian Journal of Psychology, 34*(1), 62–71.

Phillips, K. H. (1992). *Teaching kids to sing.* New York: Schocken Books.

Putz, R. V., & Tuppek, A. (1999). Evolution of the hand. *Handchirurgie, Mikrochirurgie, Plastische Chirurgie, 31*(6), 357–361.

Radman, Z. (Ed.). (2013). *The hand: An organ of the mind. What the manual tells the mental.* Cambridge, MA: MIT Press.

Reifinger, J. L., Jr. (2012). The acquisition of sight-singing skills in second-grade general music: effects of using solfege and of relating tonal patterns to songs. *Journal of Research in Music Education, 60*(1), 26–42.

Reisenweaver, A. J. (2012). Guido of Arezzo and his influence on music learning. *Musical Offerings, 3*(1), 37–59.

Repp, B. H. (2003). Rate limits in sensorimotor synchronization with auditory and visual sequences: The synchronization threshold and the benefits and costs of interval subdivision. *Journal of Motor Behaviour, 35*(4), 355–370.

Repp, B. H. (2005). Sensorimotor synchronization: A review of the tapping literature. *Psychonomic Bulletin Review, 12,* 969–992.

Rowe, M. L., Özçalişkan, Ş., & Goldin-Meadow, S. (2008). Learning words by hand: Gesture's role in predicting vocabulary development. *First Language, 28*(2), 182–199.

Sammler, D., Novembre, G., Koelsch, S., & Keller, P. E. (2013). Syntax in a pianist's hand: ERP signatures of "embodied" syntax processing in music. *Cortex, 49*(5), 1325–1339.

Simones, L., Schroeder, F., & Rodger, M. (2015). Categorizations of physical gesture in piano teaching: A preliminary enquiry. *Psychology of Music, 43*(1), 103–121.

Spilka, M. J., Steele, C. J., & Penhune, V. B. (2010). Gesture imitation in musicians and non-musicians. *Experimental Brain Research, 204*(4), 549–558.

Steeves, C. (1985). *The effect of Curwen–Kodály hand signs on pitch and interval discrimination within a Kodály curricular framework.* Unpublished master's dissertation. University of Calgary, Canada.

Strayer, H. R. (2013). From neumes to notes: The evolution of music notation. *Musical Offerings, 4*(1), 1–13.

Thaut, M. H., Kenyon, G. P., Schauer, M. L., & McIntosh, G. C. (1999). The connection between rhythmicity and brain function. *IEEE Engineering in Medicine and Biology Magazine, 18,* 101–108.

Thompson, W. (2010). *Questionnaire: Written and conducted by Helen Julia MINORS.* London: Kingston University.

Tomasello, M. (2010). *Origins of human communication.* Cambridge, MA: The MIT press.

Vets, T., Degrave, J., Nijs, L., Bressan, F., & Leman, M. (2017). PLXTRM: Prediction-Led eXtended-guitar Tool for Real-time Music applications and live performance. *Journal of New Music Research, 46*(2), 187–220.

Wilson, F. R. (1998). *The hand: How its use shapes the brain, language, and human culture.* New York, NY: Pantheon Books.

Wöllner, C. (2008). Which part of the conductor's body conveys most expressive information? A spatial occlusion approach. *Musicae Scientiae, 12*(2), 249–272.

Wolpert, D. M., Diedrichsen, J., & Flanagan, J. R. (2011). Principles of sensorimotor learning. *Nature Reviews Neuroscience, 12*(12), 739–751.

Amelynck, D., Maes, P. J., Martens, J.-P., & Leman, M. (in print). Beating-time gestures: Imitation learning for humanoid robots. *Human Robot Interactions*.

Zappa, F. (1973). Adelaide Tonight, Interview with Frank Zappa, NWS, Channel Nine, Australia 4 July, http://www.youtube.com/watch?v=sTel7GtLck4 (last accessed 14 May 2011).

Zatorre, R. J., Chen, J. L., & Penhune, V. B. (2007). When the brain plays music: Auditory-motor interactions in music perception and production. *Nature Reviews Neuroscience, 8,* 547–558.

Zinar, R. (1983). John Curwen: Teaching the tonic sol-fa method 1816–1880. *Music Educators Journal, 70*(2), 46–47.

If you have any concerns about our products,
you can contact us on
**ProductSafety@springernature.com**

In case Publisher is established outside the EU,
the EU authorized representative is:
**Springer Nature Customer Service Center GmbH
Europaplatz 3, 69115 Heidelberg, Germany**

Printed by Libri Plureos GmbH
in Hamburg, Germany